LAST

BREATH

By Peter Stark

Driving to Greenland
Last Breath

Edited by Peter Stark

Ring of Ice

LAST

BREATH

CAUTIONARY TALES
FROM THE LIMITS
OF HUMAN ENDURANCE

PETER STARK

MACMILLAN

First published 2001 by The Ballantine Publishing Group

First published in Great Britain 2002 by Macmillan
an imprint of Pan Macmillan Ltd
Pan Macmillan, 20 New Wharf Road, London N1 9RR
Basingstoke and Oxford
Associated companies throughout the world
www.panmacmillan.com

ISBN 0333 90570 9

1 3 5 7 9 8 6 4 2

A CIP catalogue record for this book is available from
the British Library.

Printed and bound in Great Britain by
Mackays of Chatham plc, Chatham, Kent

TO THOSE WHO PUSH TO THE EDGE,

AND TO THOSE WHO HELP BRING THEM BACK

CONTENTS

ACKNOWLEDGMENTS

The generosity, patience, expertise, and insight of dozens of individuals both inside and outside the medical field allowed me to write this book. With very few exceptions, everyone I contacted for help gave unstintingly of their time and knowledge; this generosity of spirit came as a great encouragement to me throughout the process of researching and writing.

I'd like to start by thanking Doug Webber, M.D., physician specializing in emergency medicine at St. Patrick's Hospital in Missoula, Montana. A climber, kayaker, scuba diver, and marathoner as well as a doctor on the front lines of the medical field, Doug guided me through many knotty questions of physiology and emergency treatment in addition to lending his outdoors expertise. I would like to thank for his psychological insights Scott Elrod, M.D., also of Missoula. I received help for many of the chapters by attending the 16th Annual Wilderness Medicine gathering in Keystone, Colorado, sponsored by the Office of Continuing Medical Education at the University of California, San Diego, School of Medicine. It was a tremendous resource to have so many experts assembled in one place, and I welcomed the opportunity to attend as a layman. Below I thank the individual physicians from the conference along with others who helped with various chapters. I have tried to render as accurately as possible what I've learned from them; I apologize to them for any lapses.

For the hypothermia chapter, which originally appeared as an article for *Outside* magazine in 1997, I had the advice of several experts on hypothermia, including Daniel Danzl, M.D.; Robert Pozos, M.D.; William Forgey, M.D.; and Cameron Bangs, M.D. Tom Bulger, M.D., also an emergency physician at St. Patrick's Hospital, helped me with details of medical procedure, and Paul Ryan of Pipestone Mountaineering, Missoula, Montana, added his expertise with outdoor equipment. Patrick Gallagher related his own harrowing encounter with hypothermia. Outdoor guide and skiing companion Skip Horner, of Skip Horner Worldwide, gave graphic details of what it's like to sleep in a tent at −48 degrees in Antarctica.

For the drowning chapter, I received help from Jerome Modell, M.D., of the University of Florida, a nationally known expert on drowning and near-drowning. Kayak instructor Kurt Doettger gave an account of his experience of near-drowning during a whitewater rescue operation in North Carolina. John Anderson and John Cox of The Trailhead outdoors store in Missoula, Montana, provided details of kayaking equipment, and Richard Gallun and Mark Wheelis lent their whitewater expertise. Jeff France of Board of Missoula described his experiences as a kayaker who has recirculated in large whitewater holes and dropped over waterfalls.

The mountain sickness chapter benefited greatly from the expertise of Peter Hackett, M.D., one of the participants in the wilderness-medicine conference and a leading researcher on the physiological effects of altitude. Also at the conference, Clifford

Zwillich, M.D., gave a helpful lecture on sleep apnea at high altitude, and Rob Roach, Ph.D., lectured on the physiology of women at high altitude. A former next-door neighbor of mine and backcountry ski partner who has gone on to become one of the Himalaya's most accomplished climbers and guides, Daniel Mazur of Himalaya Incorporated, provided many details of high-altitude climbing and its physiological effects, usually via e-mail from Tibet, Nepal, or Pakistan. Delbert Kilgore, avian physiologist at the University of Montana, helped with the physiology of birds that breathe at high altitude. Noel Ragsdale gave valuable insights on details of character, and climber Chris Brick described her own case of pulmonary edema while climbing in the Andes.

The neurofeedback descriptions and brain wave research that are included in the avalanche chapter would not have been possible without the generous assistance of freelance writer Jim Robbins of Helena, Montana, author of *A Symphony in the Brain: The Evolution of the New Brain Wave Biofeedback*, and Bernadette Pedersen, also of Helena, and her business, Brainworks Neurofeedback Services. Bernadette kindly hooked me up to a machine that read my brain waves and allowed me to try the fascinating practice of neurofeedback as I nudged my brain waves down a kind of video highway. Les Fehmi and Susan Shor-Fehmi gave their neurofeedback expertise. Scott Lewis shared his incredible story of being buried for twenty-seven minutes several feet deep in an avalanche while heli-skiing in British Columbia (he handled it much more calmly than the character described here), and Linda Parker told of her brush with an avalanche. Physical therapist Fred Lerch gave me insights on the interaction between mind and body. Beau Johnson of Board of Missoula helped with snowboarding details, and the National Weather Service office in Salt Lake City filled me in on why it snows so much in the Wasatch Mountains. Neil Beidleman, climber, engineer with Big Air Design, and lecturer at the wilderness-medicine conference, demonstrated the Black

Diamond vest designed for breathing in an avalanche. For an understanding of the physiology of avalanche victims, I am grateful to Colin K. Grissom, M.D., and Martin I. Radwin, M.D., who with their colleagues have been conducting fascinating research in the mountains of Utah by burying volunteers under the snow to simulate an avalanche and monitoring their physiology.

For the scurvy chapter, sailing companion and all-around outdoor expert Tom Duffield provided nautical accuracy about boats and the waters of Puget Sound.

For the heatstroke chapter, bicycle racer Charlie Holbrook gave an insightful account of a racer's strategy and vivid description of the strain on a racer's body when hill climbing, while Len LaBuff, owner of Open Road bicycles in Missoula, provided details of equipment. Peter Felsch, meteorologist with the National Weather Service in Missoula, and Bryan McAvoy, meteorologist with the National Weather Service in Greenville, South Carolina, helped with setting a realistic weather scenario for the chapter. P. Z. Pearce, M.D., a speaker at the wilderness-medicine conference, presented a useful lecture on heat illness.

I heard detailed, thoughtful descriptions of their long falls for the falling chapter from Otto Plattner, of Innsbruck, Austria, who tumbled far down a couloir in the Tyrolean Alps and lay in the snow for hours waiting for a rescue, and Bill Watson of Missoula, who survived, unhurt, a fall from the fourth floor of a building under construction. Climber Jack Tuholske and Susan Duffield suggested mountaineering and climbing literature for me to read, and Ann DiCesare, librarian at *Reader's Digest*, managed with amazing speed to locate the account of the tailgunner falling out of the airplane that I remembered having read in my youth, as well as the article about the Swedish hunter buried for eight days in an avalanche. For technical details of rock climbing, I am grateful to J. R. Plate of Pipestone Mountaineering and to Doug Webber, M.D., for a realistic climbing and trauma scenario. Ebo Uchimoto, professor of physics at the University of Montana, helped with the physics and calculations in the chapter, as did Jennifer Fowler, lab technician in the department. Chuck Leonard of the University of Montana's Department of Physical Therapy clarified some of the motor movements, and Charles Vogel helped with the character's financial background.

The predators chapter and description of the physiology of a box jellyfish sting benefited from the knowledge of Paul Cullen, M.D., emergency physician at Cairns Base Hospital in Cairns, Australia. The fascinating statistics on worldwide and North American land animal attacks were given in a lecture at the wilderness-medicine conference by Michael Callahan, M.D., medical director of rescue medicine at the University of Colorado, Health Sciences. Missoula writer Connie Poten, a former resident of Cairns, provided details of setting, and Nici Holt and Andy Cline told the riveting story of their excruciating entanglement with a swarm of stinging jellyfish while sea kayaking off Hawaii.

Essential details about the life and language of the Caribbean for the bends chapter came from Paul Reneau, a native of Belize who is a school counselor in Missoula. Carl Thomas, scuba instructor and employee of Gull Boats in Missoula, lent a great deal of expertise and patience in explaining details of diving equipment, planning, and techniques. Tom Neuman, M.D., of the University of California, San Diego, helped with expertise on the physiological effects of diving.

Mark Bracker, M.D., of the University of California, San Diego, School of Medicine and codirector of the wilderness-medicine conference, gave me timely and insightful comments on the physiology of malaria. I heard harrowing accounts from Bryan Di Salvatore and Jill Belsky of malaria in their traveling companions.

For the chapter on dehydration and desert survival I had the help of Ronald and Peter Kummerfeldt, instructors at the wilderness-medicine conference and survival instructors at Outdoor Safe of Colorado Springs, Colorado. Jay Christopher provided his story of dehydration while hiking in Utah.

Besides the assistance with individual chapters, I had the experience and insights of companions in recent years on various outdoor adventures of my own, among them Fred Haefele, Steve and Matt Rinella, Christopher Preston, Bill Bevis, Ted Stark, Paul Jensen, and Gray Thompson. Ira Byock, M.D., author of *Dying Well*, and Yvonne Corbeil gave me insights into the process of dying.

Vaughn Stevens and Don Spritzer of the reference desk of the Missoula Public Library always managed to find answers to my hard-to-answer questions, such as "How many stars are there in the universe?" Shaun Gant of the Mansfield Library at the University of Montana provided helpful references for several chapters, and I am grateful also to the staff of the medical library at St. Patrick's Hospital.

The idea for the book itself grew out of an article on hypothermia I wrote for *Outside* magazine. Mark Bryant, then editor of *Outside*, was instrumental in shaping what started out as a vague idea in my mind for an essay on cold and the human body. Laura Hohnhold and Gretchen Reynolds helped edit the piece and bring it to its final form. My agent, Frances Kuffel, kept up the encouragement to expand it into a book and provided a valuable sounding board throughout its conception. Jennifer Weltz of Jean V. Naggar Literary Agency has been helpful throughout. Missoula writers Bryan Di Salvatore, Steve Krauzer, Fred Haefele, and Bill Vaughn all gave me inspiration and advice at crucial moments. My father, William Stark, and my father-in-law, Wilmott Ragsdale, both writers and both great adventurers of the old school, read the chapters as I went and encouraged me by both word and example.

I've felt fortunate from the beginning that the book found a home with Ballantine. I am deeply grateful to my editor, Dan Smetanka, who has been a pleasure to work with and whose contributions to this book are beyond measure, and to his assistant, Allison Dickens, for her help and valuable input. Last, I'd like to thank my wife, Amy Ragsdale, for encouraging me to undertake the book in the first place and her unwavering support and suggestions throughout the writing of it, and my children, Molly and Skyler, who showed their own form of human endurance while waiting for me to come out of my office to play.

ARS MORIENDO:

THE CRAFT
OF DYING

I fear death.

I recall the precise moment that I grasped the utter certainty of my mortality. It didn't occur while clinging to some snowy mountainside or plunging in a kayak through Class IV whitewater—places where several victims in this book grasp the certainty of their own mortality. The moment occurred as I sat at a richly polished reading table in the hushed calm of Sanborn House, the elegant English-literature library at Dartmouth College. I don't recall what book I had open on the table before me. What I do remember is staring for a long time into the deep brown and black swirls of that rich wood grain illuminated by the reading lamp and thinking, *I'm going to die. All this will end.*

The thought was no more nor less complex than what millions—billions—of humans come to realize, sooner or later, during their short lives on earth. What surprised me then and still does was the emotional depth of that realization. I was *grief stricken* by the thought. I already mourned my own death, although I was twenty years old at the time and, barring disease or disaster, I could expect to live at least another fifty years.

I don't know precisely what triggered this realization; in retrospect, however, I find clues. Someone close to me had recently attempted suicide. Several of my friends had recently died of drug overdoses. But, despite the accumulated impact of these, I think an altogether different incident may have finally triggered my sudden realization of the inevitability of death.

I'd spent a recent spring weekend skiing at Tuckerman's Ravine, the famous thousand-foot-high headwall near the summit of New Hampshire's Mount Washington that for decades has been a mecca for snow and ice climbers—and for skiers—in search of the steep. After a few exhilarating runs down the headwall's main section, a friend and I kicked our way up the extremely steep section that is known as the Left Gully. The sun had dropped low by the time we arrived at the top, casting part of the ravine's great bowl in shadow. I proposed to my companion, who happened to be a mountain climber as well as a skier, that we ski some cleft or couloir that dropped off from where we stood. He looked over the edge, tracing with his eye the way the slope fell away into shadow and disappeared from sight, visible again a thousand feet below at the ravine's bottom. "No, we shouldn't go there," he said. "The snow down below is now in the shadow and it's freezing up hard. If we fall, we could slide a long way down into the rocks."

It hadn't occurred to me that snow conditions might change that rapidly—this was before I knew much about wilderness skiing—but what my climbing friend said made good sense. I deferred to his judgment. We skipped the steep, shadowy couloir and instead skied the larger, safer main headwall again, where the snow remained softer and no rock outcrops stood in the path of descent. About an hour later, the day's fun was winding down and people were starting the long trek down the forest trail to their cars when a helicopter suddenly buzzed into the ravine. It landed, then quickly took off again. Not until the next day did I learn what had happened. A young man skiing the steeps just in the area that I'd been eyeing had fallen on the icy, shaded snow and slid far and fast down the slick couloir into the rocks below. He died.

How easily that could have been me. It nearly *was* me but for the mountain wisdom of my friend. This incident, I think, is what triggered my revelation at the Sanborn House reading table—how close death really is, how mortality lies just a step off the curb or a wrong turn of the skis away, and just how irrevocable that ending will be.

That simple revelation occurred nearly thirty years ago. Yet still I go back to the wilderness mountains, still I love to ski the steeps, still something compels—*impassions*—me to be up there when I look out my window on a sunny spring day and see white peaks against the sky. I'm drawn by the beauty and by the silence and by the physical exertion; by the cold wind and warm sun and hot sweat, by the *exhilaration*. But another part of what compels me—a big part of it, and one that accentuates all the others—is the sense of the nearness of my own mortality that I first experienced that spring weekend at Tuckerman's Ravine. That's one premise of this book: that there are countless others who climb or ski or snowboard, who paddle whitewater, who trek in deserts, or who engage in adventures of a milder nature. Like I am, they are drawn by the beauty and silence and weather and exhilaration, and yet beneath these reasons flows an emotional undercurrent that often remains unidentified, or at least unarticulated. That is, these activities bring one nearer to an ineffable something that contains a sense of one's mortality and also reaches beyond it. Paul Bowles, writing of the compelling attraction of the Sahara Desert despite its brutal and sometimes fatal discomforts, described this something as "the absolute."

A century ago death commonly visited the bedroom next door. It was here that a relative or grandparent would die of sickness or old age, in the midst of the household. Death then was a part of everyday life. Today we don't see death in that everyday way—it is whisked away to the sanitized environment of the hospital (although fortunately the hospice movement has begun to change this). We don't talk about death. We carefully deny that death exists so close to us all, as if it has become a kind of obscenity, or, alternatively, a cardboard shoot-'em-up character in the movies. It may be our distance from death that makes us at once afraid of it, fascinated by it, and yet wanting to *understand* it. In contrast to the Western attitude toward death, the Japanese calmly prepare for their demise with a centuries-old tradition of writing in their last days or moments what are known as "death poems." The Christian tradition in medieval times placed more emphasis on familiarity with death, and scholars wrote texts such as *Ars Moriendo*, "The Craft of Dying." Throughout the world, many cultures display more familiarity with death than in the West. *The Tibetan Book of the Dead*, read aloud to the person who is dying, is an intimate portrayal of the process of death and the opportunity it presents for liberation. According to the Tibetan tradition, you must understand death intimately and prepare yourself for it in order to live a full and satisfying life.

My intention with *Last Breath* has been to produce a sort of *Ars Moriendo* for those who engage in adventurous, eminently rewarding, often risky, and sometimes fatal activities such as climbing, whitewater paddling, extreme skiing or snowboarding, and remote travel, although the themes of risk, death, and enlightenment addressed in these chapters should pertain to the many others who don't necessarily pursue these outdoor activities. Every one of us shares an inescapable fate: We are inhabitants of a body that will someday die. Moments in this book intimately follow that process as it occurs to the bodies of those who push the boundaries of outdoor activities. In each chapter I've intertwined a story of a character in a desperate outdoor situation with the physiology—the medical science—of what is happening to his or her body as this occurs. I've also tried to address some of the spiritual aspects of dying—how other cultures regard death—as well as explore, obliquely, where the scientific and the spiritual intersect in the moment of death.

During the course of research, it has never failed to amaze me how ingeniously the human body adapts to the constant changes in its external and internal environment—to changes in air pressure and oxygen, food consumption and water intake, increased physical workload, heat and cold, fear and other emotions, and blood loss. Yet that resiliency can be misleading. The human body, in many ways, is as delicate as a hothouse flower, capable of existing in only an extremely narrow band of conditions. We, as a species, are genetically adapted to a belt of tropical temperatures that prevail only around the equator; without technological innovations such as insulated clothing, furnaces, and campfires, we would quickly perish if we strayed from that narrow belt a few thousand miles north or south. While we can still roam widely within this tropical belt on the *horizontal* plane, on the *vertical* plane we can travel only a very short distance without destroying the proper functioning of our bodies. Humans cannot live on a permanent basis more than about 3.5 miles above sea level. Likewise, humans are extremely sensitive to environments that lack fresh water—deprived of fresh water, they perish within a few days, the exact number (from as few as one day to as many as ten) depending on air temperature, exertion, and other factors.

If you set the small belt that offers the precise conditions necessary for human life against all the rest of the planet—against all of the earth's surface that is desert and ice, rock and ocean—you realize exactly how limited our life-sustaining environment actually is. More important, you realize how simple it is to step beyond it, how easily we can break through the thin web of life that supports the delicate human organism. And if you set that band of life against the enormity of the solar system, or of the Milky Way galaxy, or of the unknown reaches of the universe, we as a species occupy a niche far smaller than a tiny tuft of wildflowers sheltered by a small boulder atop an otherwise barren mountain peak that harbors in that single tuft the only life on an entire planet of barren mountain peaks. This is what, at some level, science and religion both are trying to tell us— how infinitesimally small the scale of human life is against the scale of the cosmos—but at the same time they also tell us that human life forms a part of the vastness. That vastness is known by different names to different people. Some call it the expanding universe, some simply call it emptiness, and some call it God.

Last Breath's subtitle refers to the scenarios rendered in each chapter as "cautionary tales." Explorer Vitus Bering's is the only historical death described in this book. The other ten chapters portray invented characters and invented situations that approximate physiological reality as closely as possible. I've based these chapters on interviews with individuals who have survived avalanche burial, hypothermia, near-drowning, falling, pulmonary edema, and other maladies. In addition, I've interviewed specialists who treat these victims, and I have conducted extensive research in medical databases, journals, and academic as well as anecdotal sources. My major printed sources are listed after the last chapter; the acknowledgments list the many generous individuals who have helped me with the research for this book.

As I've strived to render the physiology and medical science of these deaths and near-deaths as accurately as possible, I have attempted to portray realistically the background settings and risky situations in which the victims find or place themselves. I bring a good bit of personal background to this task as a backcountry skier (and former ski racer), kayaker, and whitewater canoeist. Over the years I have traveled (usually the hard way, and usually for extended periods) to many of the remote regions portrayed here—the Himalaya and Tibet, the Arctic, the Sahara Desert, West Africa, Sumatra, China's Tiger's Leap Gorge—as well as to closer ones such as Cádiz, in the Andalusia region of Spain, where I lived for a time. I've chosen to create composite characters based on my extensive interviews and research because this gives me much greater flexibility to explore the physiological, psychological, and spiritual aspects of death than I could by tying the narrative to the experiences of a single person. My hope is that this choice finally makes for a more dramatic, comprehensive, and insightful book.

Yet these scenarios have been cautionary tales for me, too. Writing this book has, in some respects, formed a part of my ongoing personal explorations. I come from an adventurous family. My grandfather, starting early last century, loved to canoe wilderness rivers, and my father, among many other adventures, crewed in 1949 as one of the sailors on the last commercial windjammer to round Cape Horn. I inherited this legacy of adventure and had it nurtured in me when, at age four, my father and grandfather took me with them on my first overnight canoe trip. The series of adventures that started then continues to this day.

I've always felt a tension, however, about how far to push on these adventures—where, if at all, I should pull back. So many times I've found myself in a place where I feel both compelled and yet scared to go farther—whether staring from a riverbank at a set of big rapids, peering down a steep slope of powder snow that's a potential avalanche zone, or heading up with backpack on my shoulders into some mist-shrouded jungle highlands of Asia or into a small African country on the verge of revolution. I suppose this tension grows in part from the fear that I'm somehow not measuring up, that I'm letting my grandfather and father down by not pushing farther or higher or steeper, as if I'm scared of "being scared," scared of how others might judge me for it. The tension also springs from the fact that I genuinely love to do these things—ski the steeps, or paddle whitewater, or explore new terrain and unfamiliar cultures—and I hate to turn back and miss the opportunity to go on. And another part of the tension—a big part of it—is uncertainty, for uncertainty is a scary place to be.

Staring down at a steep slope or a big set of rapids, one cannot expect a definite answer to the question *What will happen?* There are only probabilities, guesses, approximations, hopes. It is here that you must rely utterly on your own judgment. To step from a daily life that is carefully bounded by laws and safety locks and guardrails into a situation where your life hinges on your own ability to assess a rapid or a slope can be both disconcerting and exhilarating. I think there is a profound desire for this kind of self-reliance among many people who live in an era when, in the Western world anyway, there is very little opportunity for it. In a difficult or risky situation in the wilderness, the total reliance on oneself and trust in one's teammates and the need for total focus—whether climbing a rock face, skiing a steep chute, or paddling a whitewater canyon—brings a crystalline awareness of the world around one and at the same time a kind of obliteration of the separateness of the self. One hears it again and again: that at moments like this the participant feels acutely alive.

There are risks, of course—risks of all sizes—and sometimes the participant pays the ultimate price for them. I've learned while writing this book that there are no sure answers, no solid black lines to demarcate caution from boldness and boldness from foolishness, or rather that those lines constantly shift depending on circumstance and individual, but this book has also helped me answer an even more important question: *Why go in the first place?* The answer, which I've tried to state through the medium of some of the characters in these chapters, has something to do with that sense of being acutely alive; it also has something to do with death.

I see how, over the years, my own attitude toward risk in the outdoors has changed. I envision this change as a progression from the ignorance of my teens, when on occasion I suspected I was skiing on an avalanche slope without being entirely sure what an avalanche slope was. My twenties, especially my early twenties, I characterize as a period of brashness; we liked to do things after late-night parties such as race iceboats at 60 miles per hour over the inky blackness of a frozen lake. As I headed toward my thirties and learned more about the outdoor world as well as about my ambitions in writing about it, brashness gave way to calculation—how large a ski jump can I handle without really hurting myself? As I've moved through my forties and toward my fifties, I find that caution has usurped calculation—or rather, I try to calculate well on the side of caution. I don't know if I'll be any more or any less cautious in the outdoors after having written this book, but I do think, paradoxically, that for all the death described in detail here, I'll fear death less—at least my own.

I fear it less than I did not only because I know more about the physiology and spiritual aspects of it than before I wrote this book, but also because I now find it easier to turn back. I no longer care as much whether I'm somehow "measuring up." Is it the self-confidence that comes with advancing age that gives me the ability to turn around, go home, and come back another day? I have small children now— that's part of it, too. I fear for them and their well-being more than my own. One's scale of adventure changes with children; it broadens and becomes simpler and at the same time infinitely more complex. With our small children, I've had adventures—and frights—in the highlands of Irian Jaya and the Sahara Desert as well as in the park across the street.

Ultimately, each person who ventures out must make his or her own decisions about how far to go and at what point to turn back. There's an old saying among prospectors who comb the hills for gold here in the American West: "Gold is where you find it." You can say the same about adventure. For that matter, you can say it about risk, about death, and about being acutely alive.

CHAPTER ONE

AS FREEZING PERSONS

PERSONS

RECOLLECT

THE SNOW:

HYPOTHERMIA

When your Jeep spins lazily off the mountain road and slams backward into a snowbank, you don't worry immediately about the cold. Your first thought is that you've just dented your bumper. Your second is that you've failed to bring a shovel. Your third is that you'll be late for dinner. Friends are expecting you at their cabin around eight for a moonlight ski, a late dinner, a sauna. Nothing can keep you from that.

Driving out of town, defroster roaring, you barely noted the bank thermometer on the town square: −27 degrees at 6:36. The radio weather report warned of a deep mass of arctic air settling over the region. The man who took your money at the Conoco station shook his head at the register and said he wouldn't be going anywhere tonight if he were you. You smiled. A little chill never hurt anybody with enough fleece and a good four-wheel drive.

But now you're stuck. Jamming the gearshift into low, you try to muscle out of the drift. The tires whine on ice-slicked snow as headlights dance on the curtain of frosted firs across the road. Shoving the lever back into park, you shoulder open the door and step from your heated capsule. Cold slaps your naked face, squeezes tears from your eyes. You check your watch: 7:18. You consult your map: A thin, switchbacking line snakes up the mountain to the penciled square that marks the cabin.

Breath rolls from you in short frosted puffs. The Jeep lies cocked sideways in the snowbank like an empty turtle shell. You think of firelight and saunas and warm food and wine. You look again at the map. It's maybe 5 or 6 miles more to that penciled square. You run that far every day before breakfast. You'll just put on your skis. No problem.

There is no precise core temperature at which the human body perishes from cold. At Dachau's cold-water immersion baths, Nazi doctors calculated death to arrive at around 77 degrees Fahrenheit. The lowest recorded core temperature in a surviving adult is 60.8 degrees. For a child it's lower. In 1994, a two-year-old girl in Saskatchewan wandered out of her house into a −40 night. She was found near her doorstep the next morning, limbs frozen solid, her core temperature 57 degrees. She lived.

Others are less fortunate, even in much milder conditions. One of Europe's worst weather disasters occurred during a 1964 competitive walk on a windy, rainy English moor; three of the racers died from hypothermia, though temperatures never fell below freezing and ranged as high as 45.

But for all scientists and statisticians now know of freezing and its physiology, no one can yet predict exactly how quickly and in whom hypothermia will strike—and whether it will kill when it does. The cold remains a mystery, more prone to fell men than women, more lethal to the thin and well muscled than to those with avoirdupois, and least forgiving to the arrogant and the unaware.

The process begins even before you leave the car, when you remove your gloves to squeeze a loose bail back into one of your ski bindings. The freezing metal bites your flesh. Your skin temperature drops.

Within a few seconds, the palms of your hands are a chilly, painful 60 degrees. Instinctively, the web of surface capillaries on your hands constrict, sending blood coursing away from your skin and deeper into your torso. Your body is allowing your fingers to chill in order to keep its vital organs warm.

You replace your gloves, noticing only that your fingers have numbed slightly. Then you kick boots into bindings and start up the road.

Were you a Norwegian fisherman or Inuit hunter, both of whom frequently work gloveless in the cold, your chilled hands would open their surface capillaries periodically to allow surges of warm blood to pass into them and maintain their flexibility. This phenomenon, known as the hunter's response, can elevate a 35-degree skin temperature to 50 degrees within seven or eight minutes.

Other human adaptations to the cold are more mysterious. Tibetan Buddhist monks can raise the skin temperature of their hands and feet by 15 degrees through meditation. Australian Aborigines, who once slept on the ground, unclothed, on near-freezing nights, would slip into a light hypothermic state, suppressing shivering until the rising sun rewarmed them.

You have no such defenses, having spent your days at a keyboard in a climate-controlled office. Only after about ten minutes of hard climbing, as your body temperature rises, does blood start seeping back into your fingers. Sweat trickles down your sternum and spine.

By now you've left the road and decided to take a shortcut up the forested mountainside to the road's next switchback. Treading slowly through deep, soft snow as the full moon hefts over a spiny ridge top, throwing silvery bands of moonlight and shadow, you think your friends were right: It's a beautiful night for skiing—though you admit, feeling the −30 degree air bite at your face, it's also cold.

After an hour, there's still no sign of the switchback, and you've begun to worry. You pause to check the map. At this moment, your core temperature reaches its high: 100.8. Climbing in deep snow, you've generated nearly ten times as much body heat as you do when you are resting.

As you step around to orient map to forest, you hear a metallic pop. You look down. The loose bail has disappeared from your binding. You lift your foot and your ski falls from your boot.

You twist on your flashlight, and its cold-weakened batteries throw a yellowish circle in the snow. It's right around here somewhere, you think, as you sift the snow through gloved fingers. Focused so intently on finding the bail, you hardly notice the frigid air pressing against your tired body and sweat-soaked clothes.

The exertion that warmed you on the way uphill now works against you: Your exercise-dilated capillaries carry the excess heat of your core to your skin, and your wet clothing dispels it rapidly into the night. The lack of insulating fat over your muscles allows the cold to creep that much closer to your warm blood.

Your temperature begins to plummet. Within seventeen minutes it reaches the normal 98.6. Then it slips below.

At 97 degrees, hunched over in your slow search, the muscles along your neck and shoulders tighten in what's known as pre-shivering muscle tone. Sensors have signaled the temperature control center in your hypothalamus, which in turn has ordered the constriction of the entire web of surface capillaries. Your hands and feet begin to ache with cold. Ignoring the pain, you dig carefully through the snow; another ten minutes pass. You know that without the bail, you're in deep trouble.

Finally, nearly forty-five minutes later, you find the bail. You even manage to pop it back into its socket and clamp your boot into the binding. But the clammy chill that started around your skin has now wrapped deep into your body's core.

At 95, you've entered the zone of mild hypothermia. You're now trembling violently as your body attains its maximum shivering response, an involuntary condition in which your muscles contract rapidly to generate additional body heat.

It was a mistake, you realize, to come out on a night this cold. You should turn back. Fishing into the front pocket of your shell parka, you fumble out the map. You consulted it to get here; it should be able to guide you back to the warm car. Your core temperature starts to slip below 95; it doesn't occur to you in your increasingly clouded and panicky mental state that you could simply follow your tracks down the way you came.

And after this long stop, the skiing itself has become more difficult. By the time you push off downhill, your muscles have cooled and tightened so dramatically that they no longer contract easily, and once contracted, they won't relax. You're locked into an ungainly, spread-armed, weak-kneed snowplow.

Still, you manage to maneuver between stands of fir, swishing down through silvery light and pools of shadow. You're too cold to think of the beautiful night or of the friends you had meant to see. You think only of the warm Jeep that waits for you somewhere at the bottom of the hill. Its gleaming shell is centered in your mind's eye as you come over the crest of a small knoll. You hear the sudden whistle of wind in your ears as you gain speed. Then, before your mind can quite process what the sight means, you notice a lump in the snow ahead.

Recognizing, slowly, the danger that you are in, you try to jam your skis to a stop. But in your panic, your balance and judgment are poor. Moments later, your ski tips plow into the buried log and you sail headfirst through the air and bellyflop into the snow.

You lie still. There's a dead silence in the forest, broken by the pumping of blood in your ears. Your ankle is throbbing with pain, and you've hit your head. You've also lost your hat and a glove. Scratchy snow is packed down your shirt. Meltwater trickles down your neck and spine.

This situation, you realize with an immediate sense of panic, is serious. Scrambling to rise, you collapse in pain, your ankle crumpling beneath you.

As you sink back into the snow, shaken, your heat begins to drain away at an alarming rate, your head alone accounting for 50 percent of the loss. The pain of the cold soon pierces your ears so sharply that you root about in the snow until you find your hat and mash it back onto your head.

But even that little activity has been exhausting. You know you should find your glove as well, and yet you're becoming too weary to feel any urgency. You decide to have a short rest before going on.

An hour passes. At one point, a stray thought says you should start being scared, but fear is a concept that floats somewhere beyond your immediate reach, like that numb hand lying naked in the snow. You've slid into the temperature range at which cold renders the enzymes in your brain less efficient. With every 1-degree drop in body temperature below 95, your cerebral metabolic rate falls off by 3 to 5 percent. When your core temperature reaches 93, amnesia nibbles at your consciousness. You check your watch: 12:58. Maybe someone will come looking for you soon. Moments later, you check again. You can't keep the numbers in your head. You'll remember little of what happens next.

Your head drops back. The snow crunches softly in your ear. In the −35 degree air, your core temperature falls about 1 degree every thirty to forty minutes, your body heat leaching out into the soft, enveloping snow. Apathy at 91 degrees. Stupor at 90.

You've now crossed the boundary into profound hypothermia. By the time your core temperature has fallen to 88 degrees, your body has abandoned the urge to warm itself by shivering. Your blood is thickening like crankcase oil in a cold engine. Your oxygen consumption, a measure of your metabolic rate, has fallen by more than a quarter. Your kidneys, however, work overtime to process the fluid overload that occurred when the blood vessels in your extremities constricted and squeezed fluids toward your center. You feel a powerful urge to urinate, the only thing you feel at all.

By 87 degrees you've lost the ability to recognize a familiar face, should one suddenly appear from the woods.

At 86 degrees, your heart, its electrical impulses hampered by chilled nerve tissues, becomes arrhythmic. It now pumps less than two-thirds the normal amount of blood. The lack of oxygen and the slowing metabolism of your brain, meanwhile, begin to trigger visual and auditory hallucinations.

You hear jingle bells. Lifting your face from your snow pillow, you realize with a surge of gladness that they're not sleigh bells; they're welcoming bells hanging from the door of your friends' cabin. You knew it had to be close by. The jingling is the sound of the cabin door opening, just through the fir trees.

Attempting to stand, you collapse in a tangle of skis and poles. That's okay. You can crawl. It's so close.

Hours later, or maybe it's minutes, you realize the cabin still sits beyond the grove of trees. You've crawled only a few feet. The light on your wristwatch pulses in the darkness: 5:20. Exhausted, you decide to rest your head for a moment.

When you lift it again, you're inside, lying on the floor before the woodstove. The fire throws off a red glow. First it's warm; then it's hot; then it's searing your flesh. Your clothing has caught fire.

At 85 degrees, those freezing to death, in a strange, anguished paroxysm, often rip off their clothes. This phenomenon, known as paradoxical undressing, is common enough that urban hypothermia

victims are sometimes initially diagnosed as victims of sexual assault.
Though researchers are uncertain of the cause, the most logical explanation is that shortly before loss of consciousness, the constricted
blood vessels near the body's surface suddenly dilate and produce a
sensation of extreme heat against the skin.

All you know is that you're burning. You claw off your shell and
pile sweater and fling them away.

But then, in a final moment of clarity, you realize there's no
stove, no cabin, no friends. You're lying alone in the bitter cold,
naked from the waist up. You grasp your terrible misunderstanding,
a whole series of misunderstandings, like a dream ratcheting into
wrongness. You've shed your clothes, your car, your oil-heated house
in town. Without this ingenious technology, you're simply a delicate,
tropical organism whose range is restricted to a narrow sunlit band
that girds the earth at the equator.

And you've now ventured way beyond it.

There's an adage about hypothermia: "You aren't dead until you're
warm and dead." This points up the importance of attempting to
warm and revive seemingly dead victims, for despite all appearances
to the contrary, they might still retain the spark of life.

At about 6:00 the next morning, his friends, having discovered
the stalled Jeep, find him, still huddled inches from the buried log,
his gloveless hand shoved into his armpit. The flesh of his limbs is
waxy and stiff as old putty, his pulse nonexistent, his pupils unre-
sponsive to light. Dead.

But those who understand cold know that even as it deadens, it
offers perverse salvation. Heat is a presence: the rapid vibrating of
molecules. Cold is an absence: the damping of the vibrations. At absolute zero, −459.67 degrees Fahrenheit, molecular motion ceases al-

together. It is this slowing that converts gases to liquids, liquids to solids, and renders solids harder. It slows bacterial growth and chemical reactions. In the human body, cold shuts down metabolism. The lungs take in less oxygen, the heart pumps less blood. Under normal temperatures, this would produce brain damage. But the chilled brain, having slowed its own metabolism, needs far less oxygen-rich blood and can, under the right circumstances, survive intact.

Setting her ear to his chest, one of his rescuers listens intently. Seconds pass. Then, faintly, she hears a tiny sound—a single thump, so slight that it might be the sound of her own blood. She presses her ear harder to the cold flesh. Another faint thump, then another.

The slowing that accompanies freezing is, in its way, so beneficial that it is even induced at times. Cardiac surgeons often use deep chilling to slow a patient's metabolism in preparation for surgical procedures. In this state of near suspension, the patient's blood flows slowly, his heart rarely beats—or, in the case of those on heart-lung machines, doesn't beat at all; death seems near. But carefully monitored, a patient can remain in this cold stasis, undamaged, for hours.

The rescuers quickly wrap their friend's naked torso with a spare parka, his hands with mittens, his entire body with a bivvy sack. They brush snow from his pasty, frozen face. Then one snakes down through the forest to the nearest cabin. The others, left in the predawn darkness, huddle against him as silence closes around them. For a moment, the woman imagines she can hear the scurrying, breathing, snoring of a world of creatures that have taken cover this frigid night beneath the thick quilt of snow.

With a "one, two, three," the doctor and nurses slide the man's stiff, curled form onto a table fitted with a mattress filled with warm water that will be regularly reheated. They'd been warned that they had a profound hypothermia case coming in. Usually such victims can be straightened from their tortured fetal positions. This one can't.

Technicians scissor with stainless-steel shears at the man's urine-soaked long underwear and shell pants, frozen together like corrugated cardboard. They attach heart-monitor electrodes to his chest and insert a low-temperature electronic thermometer into his rectum. Digital readings flash: 24 beats per minute and a core temperature of 79.2 degrees.

The doctor shakes his head. He can't remember seeing numbers so low. He's not quite sure how to revive this man without killing him.

In fact, many hypothermia victims die each year in the process of being rescued. In what's called rewarming shock, the constricted capillaries reopen almost all at once, causing a sudden drop in blood pressure. The slightest movement can send a victim's heart muscle into wild spasms of ventricular fibrillation. In 1980, sixteen ship-wrecked Danish fishermen were hauled to safety after an hour and a half in the frigid North Sea. They then walked across the deck of the rescue ship, stepped below for a hot drink, and dropped dead, all sixteen of them.

"Seventy-eight-point-nine," a technician calls out. "That's three-tenths down."

The patient is now experiencing after-drop, in which residual cold close to the body's surface continues to cool the core even after the victim is removed from the outdoors.

The doctor rapidly issues orders to his staff: intravenous administration of warm saline, the bag first heated in the microwave to 110 degrees. Elevating the core temperature of an average-size male 1 degree requires adding about 60 kilocalories of heat. A kilocalorie is the amount of heat needed to raise the temperature of 1 liter of water 1 degree Celsius. Since a quart of hot soup at 140 degrees offers about 30 kilocalories, the patient curled on the table would need to consume 40 quarts of chicken broth to push his core temperature up to normal. Even the warm saline, infused directly into his blood, will add only 30 kilocalories.

Ideally, the doctor would have access to a cardiopulmonary bypass machine, with which he could pump out the victim's blood, warm and oxygenate it, and pump it back in again, safely raising the core temperature as much as 1 degree every three minutes. But such machines are rarely available outside major urban hospitals. Here, without such equipment, the doctor must rely on other options.

"Get me the peritoneal dialysis setup," he calls out.

Moments later, he's sliding a large catheter into an incision in the man's abdominal cavity. Warm fluid begins to flow from a suspended bag, washing through his abdomen, and draining out through another catheter placed in another incision. Prosaically, this lavage operates much like a car radiator in reverse: The solution warms the internal organs, and the warm blood in the organs is then pumped by the heart throughout the body. A catheter is inserted into his bladder, and warm fluid from another suspended bag flows into it.

The patient's stiff limbs begin to relax. His pulse edges up. But even so, the jagged line of his heartbeat flashing across the EKG screen shows the curious dip known as a J wave, common to hypothermia patients.

"Be ready to defibrillate," the doctor warns the EMTs. Yet he is aware how rare it is at such low core temperatures that an electrical shock will restore normal heart rhythm.

For another hour, nurses and EMTs hover around the edges of the table where the patient lies centered in a warm pool of light. They check his heart. They check the heat of the mattress beneath him. They whisper to one another about the foolishness of having gone out alone tonight.

And slowly the patient responds. Another liter of saline is added to the IV. The man's blood pressure remains far too low, brought down by the blood flowing out to the fast-opening capillaries of his limbs. Fluid lost through perspiration and urination has reduced his blood volume. But every fifteen or twenty minutes, his temperature rises another degree. The immediate danger of cardiac fibrillation lessens as the heart and thinning blood warms. Frostbite could still cost him fingers or an earlobe. But he appears to have beaten back the worst of the frigidity.

For the next half hour, an EMT quietly calls the readouts of the thermometer, a mantra that marks the progress of this cold-blooded proto-organism toward a state of warmer, higher consciousness.

From somewhere far away in the immense, cold darkness, you hear a faint, insistent hum. Quickly it mushrooms into a ball of sound, like a planet rushing toward you, and then it becomes a stream of words.

A voice is calling your name.

You don't want to open your eyes. You sense heat and light playing against your eyelids, but beneath their warm dance a chill wells up inside you from the sunless ocean bottoms and the farthest depths of space. You are too tired even to shiver. You want only to sleep.

"Can you hear me?"

You force open your eyes. Lights glare overhead. Around the lights faces hover atop uniformed bodies. You try to think: You've been away a very long time, but where have you been?

"You're at the hospital. You got caught in the cold."

You try to nod. Your neck muscles feel rusted shut, unused for years. They respond to your command with only a slight twitch.

"You'll probably have amnesia," the voice says.

You remember the moon shining over the spiky ridge top and skiing up toward it, toward someplace warm beneath the frozen moon. After that, nothing—only that immense coldness lodged inside you.

"We're trying to get a little warmth back into you," the voice says.

You'd nod if you could. But you can't move. All you can feel is throbbing discomfort everywhere. Glancing down to where the pain is most biting, you notice blisters filled with clear fluid dotting your fingers, once gloveless in the snow. During the long, cold hours the tissue froze and ice crystals formed in the tiny spaces between your cells, sucking water from them, blocking the blood supply. You stare at them absently.

"I think they'll be fine," a voice from overhead says. "The damage looks superficial. We expect that the blisters will break in a week or so, and the tissue should revive after that."

If not, you know that your fingers will eventually turn black, the color of bloodless, dead tissue. And then they will be amputated.

But worry slips from you as another wave of exhaustion sweeps in. Slowly you drift off, dreaming of warmth, of tropical ocean wavelets breaking across your chest, of warm sand beneath you.

Hours later, still logy and numb, you surface, as if from deep under water. A warm tide seems to be flooding your midsection. Focusing your eyes down there with difficulty, you see tubes running into you, their heat mingling with your abdomen's depthless cold like a churned-up river. You follow the tubes to the bag that hangs suspended beneath the electric light.

You begin to understand. The bag contains all that you had so nearly lost. These people huddled around you have brought you sunlight and warmth, things you once so cavalierly dismissed as constant, available, yours, summoned by the simple twisting of a knob or tossing on of a layer.

But in the hours since you last believed that, you've traveled to a place where there is no sun. You've seen that in the infinite reaches of the universe, heat is as glorious and ephemeral as the light of the stars. Heat exists only where matter exists, where particles can vibrate and jump. In the infinite winter of space, heat is tiny; it is the cold that is huge.

Someone speaks. Your eyes move from bright lights to shadowy forms in the dim outer reaches of the room. You recognize the voice of one of the friends you set out to visit, so long ago now. She's smiling down at you crookedly.

"It's cold out there," she says. "Isn't it?"

CHAPTER TWO

A RIVER OF

ONE'S OWN:

DROWNING

Under heaven nothing is more soft and yielding than water
Yet for attacking the solid and strong, nothing is better.
TAO TE CHING, SIXTH CENTURY B.C.

Matt knew in the first instant that he'd entered rapids far more powerful than any he'd paddled before. From the calm eddy near the shore he gave three quick strokes to drive his kayak into the mainstream. His boat's bow was seized and jerked downstream by the river that roller-coastered through the canyon, the sudden acceleration snapping back his head and torso in the cockpit as if he'd stomped the gas pedal of a race car. Instinctively he pressed his paddle blade facedown in the water to brace his boat upright. Leaning on the blade as he crested the first great, rolling wave, a single thought whisked through Matt's mind: *Holy shit!*

To Matt's kayaker friends on the shore, the river's cascade of foam and noise and spray resembled a freight train, flying by so close and fast and loud you couldn't make out the rail company's name on an individual car without quickly swinging your eyes and head to follow its motion. The entire volume of the Yangtze River cascaded over the edge of the Tibetan Plateau, the "Roof of the World," and spilled down a 500-mile-long series of gorges like a cloudburst down a giant drainpipe. Halfway down, this giant drainpipe was blocked by the Jade Dragon Mountains, a steep, beautiful, snow-capped range that formed an 18,000-foot-high dam. Eons ago, an earthquake had snapped the mountains' spine with a 2-mile-deep fracture—so narrow and so deep that the Chinese knew it as Tiger's Leap Gorge, for the hunted tiger that once had supposedly leapt across it. Here the world's fourth-longest river literally tipped sideways to squeeze through this crack in the mountains in a roaring chaos of water seeking the quickest path to the sea.

No one had ever run the gorge in a kayak. No one had even *tried*, so daunting and remote was this stretch of river. At great expense and effort, Matt and his friends had traveled to Yunnan Province to attempt the gorge's first kayak descent—a feat that, if successful, would claim a small but significant place in kayaking history. For three days they'd camped on a broad shelf of rock carved from the gorge's wall to scout the rapids, scanning the river through binoculars and scrambling over the boulders along the shore. The power and noise and size of the water was far larger in real life than anything they'd imagined from the warm safety of home, and Matt, finally, was the only one willing to attempt it.

"I don't know, Matt," one of the group had cautioned around the campfire. "You're the best paddler here, but did you see the *size* of those holes? They could swallow an entire cargo raft, not to mention your little kayak."

Matt stared into the flames. He could hear the distant roar of the water from deep in the canyon, ululating on the night breeze like the ebb and flow of far-off surf. In his mind's eye, he saw the holes, where enormous boulders had fallen from the gorge's walls into midstream and now stood as obstacles to the river's flow. As the current flowed up and over a boulder, it fell down the far side and formed a depression in the river's surface—a "hole"—always on the downstream side. Rivers constantly tried to "fill" their holes by pouring in water from just downstream of the hole or from the sides, creating a foaming, spinning drain in the river's smooth flow.

You could find tiny holes in the smallest brooks, their little mustaches of whitewater just downstream of pebbles. On a big, steep river, the water thundered into a hole with such force that, should you inadvertently tip your craft into one, it could pull you, your life jacket, and your kayak down into it. It could bring you up to the surface again, pull you down again, bring you up—endlessly recirculating you as the water spins through the hole. This was known as a "keeper." From a keeper, there is only one exit: straight down. You would let the water pull you deep, and then—against all instinct, and while the oxygen disappears from your burning lungs— you swim even deeper, until finally the force of the down-pouring water spends itself on the river bottom and discharges you downstream. Then you swim to the surface. Assuming, of course, you're still alive.

The flames bent in the breeze, and the distant roar of the rapids sounded as if it had gone underground, then emerged again. Matt had missed his line and dumped his kayak into holes before, lots of times, although the holes he'd seen the last three days were bigger— *much* bigger—than any he'd encountered in all his years kayaking. He'd started paddling with his friends the summer after their freshman year in college. When fall had come that year, he just couldn't walk away, as they managed to do, from the searing intensity and delicate finesse of a ride through whitewater and settle back into a classroom.

"What am I *doing* here?" he asked himself after the first twenty minutes of the first class, as the professor droned on about the fundamentals of accounting. When the bell rang at the end of the hour he walked to the bookstore and returned the fat accounting textbook for cash. He never looked back.

Since then, he'd scraped by with whatever decent seasonal job he could find—laying asphalt, pouring concrete, building swimming pools—and spent all his spare time seeking out whitewater rivers. He organized a big trip once a year with his old kayaking buddies, and it had been his suggestion that they try Tiger's Leap Gorge. He was surprised at first that they'd agreed to his proposal; either they didn't comprehend quite what a fearsome stretch of water it was, or they knew they could always bow out of the paddling. For Matt, however, it was much more complicated. They would go home and make their marks in their professional fields—their software designing and their company managing—and simply chalk up a failed expedition as a nice trip to an exotic place. But kayaking was what he *did*. It was how he thought of himself. He wasn't a company manager or a software designer or a securities trader, he was a kayaker. This, right here—this gorge, this first descent, this moment—was his.

Matt looked up. The firelight reflected in their eyes as they watched him. "I've worked for two years framing houses to make enough money for this trip," he said. "I'm not going home without at least trying. Besides, I'm not planning on dumping into one of those big holes. And if I do—well, I guess I'll just have to hang on, go deep, and wait till it kicks me out below."

No one spoke. They looked down into the hissing coals and up at each other, silently asking, *Should we try to stop him?*

"Have you ever read Lao Tzu—the Tao Te Ching?" Matt asked suddenly, looking around at the silent, firelit faces. After he quit college, he'd made a point of continuing to read, as if to compensate for his lack of studies, and tightly gripped his discoveries with the passionate single-mindedness of the self-educated. "It's a lot about water," he told them. "The way it moves and how it flows around obstacles like rocks and a lot about how this is the way you should live your life. That you have to yield to something in order to overcome it. That's how a paddler has to think. Especially in a big hole. You can't fight the hydraulics. You have to give yourself over to it in order to be let go."

His friends looked at each other again across the flames. "You're talking Class Six rapids in that gorge, Matt," said one. "Wanting to run it as badly as you do doesn't sound like 'yielding' to me."

"Well, I'm going to do it anyway," Matt said abruptly. "You can either help me, or I'll do it alone."

After a breakfast the next morning of green tea and Chinese noodles, Matt and his friends hauled his equipment down to a slab of rock near the water. Though the kayakers each usually looked after his or her own equipment and preparations, they all now helped Matt dress for the run, attending to him like squires preparing a knight for battle. They held open the back of his full-length Gore-Tex dry suit as he climbed into it, zipped it up the back, and fussed with the rubber gaskets around his wrists, ankles, and neck, designed to seal the water out, tightening down each with a Velcro overlay strap. They knelt before him and helped him slip on his river shoes, designed with mesh uppers to drain water, and rubber soles stiff enough to push away from submerged boulders should he have to swim from his kayak, but sticky enough to scale a canyon wall if that was his only escape from the gorge.

As his friends cinched straps and sealed Velcro and zipped zippers, Matt's body was already shifting into battle mode. Triggered by the fearsome pounding of the river and the task at hand, his brain signaled his adrenal glands, perched atop each kidney, to release. He felt a fluttery quickness spread in his torso as they poured adrenaline and noradrenaline into his heart, which pumped it to his muscles and organs, where the volatile molecules contracted the fibers to prepare him for either fight or flight when he met his foe. His 35 feet of small intestine shut down, his anus snapped closed, his liver dumped out sugar into his bloodstream to fuel his muscles, and his heartbeat suddenly doubled, jumping from its resting rate of 60 beats per minute to 120. The adrenaline hit the muscles around his chest and diaphragm, and his breathing surged and deepened from 12 shallow breaths per minute to 20 deep ones. His body was taking on oxygen as if it knew, more than he, just how badly he'd be needing it.

His friends helped strap on his high-flotation, bright yellow life vest, easy to spot should he be trapped underwater, and equipped with a built-in harness to which they could clip a carabiner and rescue rope, attaching it to a pulley system rigged on shore to hoist him out if he or his kayak was pinned by the current against a log or boulder. Finally, like squires lowering a battle helmet over the knight's head, they fitted his Kevlar paddling helmet and strapped it beneath his chin. This would protect his skull from collisions with underwater rocks if his kayak tipped and he floated along upside-down underwater attempting to right it with that quick flick of his paddle and hips known as an Eskimo roll.

"Do you want your face guard?" one of his friends asked, holding up a meshwork of metal that fitted to the front of his helmet to protect his face from contact with submerged rocks.

Matt brushed it away. "I'd rather be able to see where I'm going."

Carrying his bright red polyethylene kayak and carbon-fiber paddle with the translucent yellow Kevlar blades, the group shuffled over the ancient rock of the gorge to the slab's edge, where it dropped abruptly 5 or 6 feet to the river's surface. Just upstream from where they stood, a point of rock projected from the shore into the roaring current and formed a back eddy—a quieter, sheltered piece of water where the current actually travels upstream like the air swirling in behind a windbreak. Matt snuggled his legs down inside his kayak's fat, stubby hull, padded exactly to fit his body, and stretched his spray squirt around the cockpit to prevent water from leaking in. The kayak's design was known as a "creek boat"—preferred by expert kayakers for running steep and very technical whitewater, stable yet quick-turning, and its hull designed to cushion the impact of kayakers free-falling over waterfalls and landing their boats with the boat bottoms flat to the river in the maneuver known as a "boof."

Matt loved the rhythm of whitewater kayaking: first the mist-soaked concentration as he clambered along the shore scouting for a perfect, clean line through the puzzle of rocks and crosscurrents and holes that formed a rapid; then assembling in his mind a series of moves like a choreographed dance. The run as he wished it to happen etched and rehearsed and rerehearsed in his memory, there followed the preparations of dressing while the tension built inside him; then the incredible intensity and focus and exertion of the run itself, and, if it went well, the final release in a gentle pool below the rapids when serenity and warmth and relief wrapped his whole being. If big-water kayaking resembled some high-speed, high-stakes game of chess, where the moves were memorized beforehand, running a section as powerful and untried as Tiger's Leap Gorge was like playing against a grand master who had never lost a game.

Matt's friends strapped on their own life vests and spread out along the ledges and rocks along the shore, each carrying a bag that contained a floating polypropylene throw rope that they could fling to Matt in case he needed help. Sealed into the cockpit, Matt used his hands like a seal uses flippers to maneuver his kayak to the very lip of the rock shelf, took up his paddle, waved to his friends, and gave a final jerk of his hips. The kayak tilted off the ledge and boofed down into the eddy with a splash.

With three quick strokes, he launched out into the roller-coastering surge of the main current, vaguely sensing his friends' presence sweeping past him on shore, blue and red smudges of rain parkas and life vests in the corner of his eye, hearing shouts growing faint in the roar: "Go, Matt! Go!"

Matt held his brace, leaning on his paddle, over the first big rolling wave, down into the trough, and up the second, ready to make his move. Some of the toughest maneuvering came at the beginning of the run—an S-curve between three big boulders, and then a hard pull to the left to avoid where the river cascaded over a chunk of rock as large as a house that sat to the right of the river's center. Just downstream of the rock loomed a hole as big as a five-car garage. That's what he had to avoid.

Spotting the first boulder of the S-turn, Matt paddled hard right, back left, then right again, weaving his way through the obstacles. To his friends onshore it appeared his red helmet and whirling yellow paddle blades were some child's toy bouncing down an enormous foaming set of stairs.

Then the House Rock. *Left! Left! Left!* When he'd scouted it from shore, Matt hadn't noticed how profoundly the current shifted here to the right—the opposite of the direction he needed to go. He pulled with all his strength, feeling the flex of his paddle shaft, the clench and release of his biceps, the wheeze of his lungs, the dig of the blades. He was fighting the entire force of the current that slid sideways in a bubbling, swirling sheet like a glacier that had come to life, shoving him inexorably across the channel toward the wrong shore, toward House Rock. Matt's cardiovascular system was now in overdrive. His breathing rate jumped to 35 breaths a minute, near the maximum for efficient breathing, taking in 130 liters of air—about 35 gallons—each minute. His heart beat at 190 beats per minute and pumped nearly 30 liters of blood, or over 7 gallons each minute, enough to fill a bathtub up to the overflow drain in four minutes. This river of blood, the amount a trained athlete's heart puts out near maximum performance, is what it took to supply Matt's muscles with enough oxygen to paddle Tiger's Leap Gorge.

His paddle blades whirled and pulled against the flow, straining left, and his little cork of a craft was still being swept to the right. Thirty yards away he could see how the moving sheet of water arched in a huge glassy bulge over House Rock, like a crystal dome. *Pull. Pull. Pull.* He could feel the sharp cut of the breeze on his wet face as the current accelerated toward the rock. He had to pull harder. His kayak was spinning sideways. He was being whisked up that glassy bulge. He could hear the thundering roar of the hole beyond the bulge. He braced his paddle to try to straighten his kayak. He saw a maelstrom of white and noise far below him. He was being tossed into the air. Then he was falling upside down, vainly paddling through the empty air, with the kayak still attached to his legs trailing behind him like some molded-polyethylene kite tail, as he plunged down into the hole's spinning eye.

His friends on shore couldn't see what happened. They watched Matt shoot up on top of an enormous rooster tail of water as if launching from a ski jump, his blades windmilling in the air, and then he was gone.

One of the sea's most exceptional diving mammals, the white whale *Delphinapterus leucas* (or beluga), can hold its breath for seventeen minutes and plunge to depths of half a mile under the ocean's surface. The elephant seal can remain submerged for up to two hours. Other seals can dive nearly a mile deep. These deep-diving marine mammals manage their stunning feats by storing oxygen in tissues and blood while breathing on the surface, and drawing on their reserves underwater. Some have respiratory systems that allow their lungs to collapse under the pressure of extreme depths, and some drop their heart and metabolic rates on extended dives to curb oxygen use. It is believed that deep-diving penguins rely on anaerobic metabolism—metabolism without oxygen—to power their swimming muscles while underwater, while ducks allow their brains to cool to temperatures low enough to prevent brain damage from lack of oxygen.

Humans, however, are very different. Unlike the marine mammals, one unfortunate side effect of our evolution as a species is that as our human progenitors climbed from the warm seas and began to walk about on dry land, they retained almost zero capacity to warehouse oxygen. We as a species possess very little capacity to hold our breath. For precisely this reason, drowning is the third most common cause of death from accidental injury in the United States, at 8,000 deaths per year. In Europe, 35,000 people drown annually, and worldwide nearly half a million people perish, placing drowning in fourth place in deaths from injuries—just ahead of falls and just behind murder and violence.[1]

More than many forms of death, drowning retains a literary quality, as if the act of submersion itself connotes surrender, submission to something greater, or, among the despairing, the abandoning of all hope. Having used water imagery in much of her fiction, Virginia Woolf, fearing another bout of madness in the spring of 1941 as German bombs dropped on England, loaded the pockets of her heavy fur coat with stones and waded out into a favorite stretch of the River Ouse. The first wife of the romantic poet Percy Bysshe Shelley drowned herself after Shelley took up with a sixteen-year-old girl, and Shelley himself drowned a few years later when his small yacht, *Ariel,* capsized off the coast of Tuscany in 1822.

For centuries, drowning has loomed in the popular imagination as a gruesome but often-preventable form of death. Europe and America's original humane societies, founded in the late 1700s, were chartered not for the humane treatment of animals but for reviving nearly drowned humans—such as London's Society for the Recovery of Persons Apparently Drowned. It was later renamed the Royal Humane Society for the Apparently Dead, to include other types of what was called "sudden death," such as falls and lightning strikes. The London and Amsterdam societies alone were said to have saved over a thousand lives in a twenty-year period.

Their detailed case reports describe how to perform CPR techniques developed in the late eighteenth century that appear

---

1. *World Health Report 2000: Annex Table 3: Deaths by cause, sex and mortality stratum in WHO Regions, estimates for 1999.* Geneva, Switzerland: World Health Organization, 2000. Traffic accidents and suicides are the world's leading causes of death from injuries.

startlingly modern: mouth-to-mouth resuscitation ("the medium of a handkerchief or cloth may be used to render the operation less indelicate"), the use of tubes to keep the breathing passages clear, and the use of bellows, which all resuscitation teams were recommended to carry, to assist the ventilation of the lungs. Rescuers discovered that electric shock could stimulate a stopped heart back into action. In 1819 Giovanni Aldini, nephew of the famed researcher of electricity Luigi Galvani, wrote a book titled *The Application of Galvanism for Medical Purposes, Particularly in Cases of Suspended Animation*. The Humane Society of New York concurred that electricity was indeed a "most powerful agent, a very proper remedy," recommending that "strong sparks . . . should be drawn from the left side over the heart." It appears from her own introduction to the book that this use of electricity to revive the near-dead was inspiration for Mary Wollstonecraft Shelley, wife of the soon-to-be-drowned poet, for her 1818 novel of terror where the monster comes to life with the application of a bolt of lightning, *Frankenstein*.

Dr. Frankenstein's fame notwithstanding, many of these techniques were abandoned or lost and not rediscovered and refined by researchers until the 1950s. Today they are used routinely. One difference, however, is that those eighteenth-century victims succumbed in lakes, rivers, and the sea, while in the United States about half of today's victims (and many in Australia) drown in swimming pools—especially toddlers who fall into private pools. Unlike a few centuries ago, drownings in the sea are comparatively rare except in cold-water fishing communities. Drownings among whitewater boaters—kayakers, canoeists, or rafters who are paddling in rapids—also make up only a small fraction of all drowning deaths, under fifty annually in the United States according to whitewater accident reports, the number having climbed as more paddlers pursue ever steeper and bigger descents.

Whatever the specific circumstances—a sinking ship, a swimming pool, a rock-laden fur coat, a kayak in Class V rapids—drowning is the archetype for all human deaths. At the very end almost everyone—heart patient, cancer victim, the very aged—perishes from the very cause so dramatically embodied in drowning: lack of oxygen to the brain. In the case of these illnesses, it can take days or weeks or years to reach that final moment. But drowning accelerates the process to a horrifying pace: the average person can hold his or her breath for about a minute and a half before inhaling or blacking out. In a victim on dry land whose breathing has stopped, brain damage starts to occur after about four minutes, and doctors consider the chances of brain recovery virtually nil after ten minutes. Underwater, however, especially cold water, these times can attenuate. One study of fifty-seven near-drowning survivors showed they were totally immersed for an estimated average of eleven and a half minutes. This is still a very short time.

0 minutes, 03 seconds: *Get a few big breaths!* That's Matt's only thought as he wheels upside down through the air over House Rock. He has no time to think, no time to be scared, no time except for the most instinctive reactions. He manages to suck in just short of 5 liters—a little over 5 quarts—of air in his last breath. This is nearly his vital capacity, or the maximum amount his lungs can take in with a single breath, which ranges among individuals from 2 to 6 liters. The air Matt pulls in, just like every other breath of the earth's atmosphere, is composed of four-fifths nitrogen and one-fifth oxygen. This gives him 1 liter of precious oxygen trapped in his lungs—about a quart—to keep the spark of life burning for his journey beneath the surface.

He plunges into the hole headfirst. He is instantly ripped out of his kayak. The water snaps his paddle shaft in two and tears it from his grip, strips the contact lenses from his corneas, pulls the helmet off his head, and yanks at the straps of his life vest as if some great animal were shaking him in its teeth. He feels himself plunging down through layers of water as if slicing through some giant layer cake: first whitish foam, then foam churned with green water, then invisible clouds of dark-green water buffeting him silently as he's pulled toward the river bottom. Far above him he can hear the deep bass rumble of the water thundering over the rock.

Profound changes in his physiology occur even in these first few seconds. All humans, to a greater or lesser degree, retain a ghostly genetic inheritance of the marine mammals' diving response, which flips one's metabolic switch from fast to slow in order to conserve oxygen while underwater. Simply by dipping your face in ice water, you can drop your heart rate from 70 to 45 beats per minute, and some athletes can lower their pulse rate with ice-water facials to under 6 beats per minute. As his heart rate drops, Matt's veins and arteries in his arms and legs constrict to channel his remaining oxygenated blood to his brain, heart, and other vital organs. Like a seal or whale, his body strains to conserve that single liter—that 1,000 milliliters—of oxygen.

o minutes, 12 seconds (825 milliliters of oxygen remaining): Matt stays down, holding his breath. Already he can feel the strain in his lungs. Carbon dioxide—what his cells expel as a waste product after they've burned fuel—is building in his muscles and organs and is taken away by his bloodstream to his lungs. Extremely precise sensors in his brain "taste" the slight acidic content of his blood from the buildup of carbon dioxide, and signal his lungs—urgently—to exhale it. It's this carbon dioxide buildup, and not lack of oxygen, that largely triggers his urge to breathe. For this reason, underwater distance swimmers sometimes hyperventilate beforehand to partially cleanse their blood of carbon dioxide, helping them short-circuit the body's urge to breathe. "I felt like I could swim forever without a breath," is how some of them have described their state shortly before they blacked out and had to be dragged from the bottom of the pool. Matt managed a good breath before going under, but he didn't hyperventilate and hasn't totally overridden his breathing urge. He feels an intensifying pressure inside his chest, almost like an inflating balloon. *Breathe out!* his body is telling him. *Expel that carbon diox-ide! Breathe out!*

Hang on, he's telling himself, marshaling all his will to fight that expanding pain in his chest. *Don't panic,* he tells himself. *You've been here before.* It's a giant Maytag, as the kayakers say—a hole that end-lessly recirculates whatever it's caught. He can feel himself being lifted toward the surface by the billows of black current welling up from down deep and, as he rises, slammed back down toward the bottom by sledgehammer blasts of whitewater that have cascaded over the rock. His lungs scream for him to break to the surface, to struggle wildly to the top. *Stay down!* he tells himself. *Don't fight the water! Go deeper!* Overcoming every human instinct to go up, he kicks and strokes downward, bumping and scraping down along the sub-merged face of House Rock, which slides past him like the wall of an elevator shaft.

o minutes, 22 seconds (640 milliliters of oxygen remaining): He can't hear the profound thunder above him anymore, and the water stays consistently dark. He senses he's been flushed out of the hole's Maytag cycle and sent downstream. But still-powerful underwater crosscurrents flail him about, his arms and legs whipped around like lengths of wet tissue paper as he tries to relax and let the water take him until it slows a bit and he can surface. Under the buffeting of the currents, his life jacket feels as if it has all the buoyancy of a soaked sweatshirt. It's as if he's swum into a herd of huge under-water animals whose fins brush and knock him about.

Meanwhile, remarkable events unfold inside his lungs. Matt's precious liter of oxygen fills three hundred million tiny sacs in each lung called alveoli, which, if spread out, would cover an area half the size of a tennis court. These sacs are enmeshed with blood-filled capillaries so fine and extensive that one physiologist likens them to spreading a single glassful of blood so thinly, it covers the entire sur-face of that half-court. What's more, this blood is squeegeed up from the tennis court and a new glassful spread over its surface more than once each second as his heart pumps blood through his lungs.

In that fraction of a second each glassful of blood unloads its waste carbon dioxide and dumps it into his lungs and at the same time loads up with a fresh cargo of oxygen in the process known as gas exchange.

o minutes, 28 seconds (490 milliliters of oxygen remaining): The buffeting along the river bottom gentles for a moment. He's entered a calmer pool, a slack spot in the current's flow. *This is it*, he thinks. *Make your move.* He kicks and strokes upward, trying not to flail, conserving his body's resources. As he slices upward, the water color lightens—dark green to pale green to swirling foam. The bubbles spin around his head and in his ears—the crackling, breathy sound of the surface, of air, of life. He's about to break the surface. This is it. He'll suck in a huge lungful of fresh air. But the current abruptly lifts his body, tosses it down. He tumbles over a submerged boulder. Another hole. Down, down, down he goes again through the layers of light. It's black. He can't breathe. He's tumbling over and over. He has to get out, to fight, claw, scratch his way out. *No!* he tells himself. *Yield to it! Yield, and overcome!*

o minutes, 37 seconds (415 milliliters of oxygen remaining): Already his consciousness has begun to shrink from lack of oxygen, like a globe of light that slowly contracts and dims. His oxygen delivery and carbon dioxide monitoring systems are on red alert. Instead of a rich, oxygenated red, his blood has begun to turn bluish. His heart thumps and his ears ring loudly with its thumping. His thoughts turn slower and slower beneath the water: *You're going to die if you stay here any longer.*

Part of him wants to panic, but he can't quite manage it, as if the panic itself occurs in slow motion. He kicks and pulls for the surface. His actions seem strangely ineffective, dreamlike. He is suffering the symptoms of hypoxia—lack of oxygen reaching the body's tissues. Distantly he can feel his arm and leg muscles burn, like the far-off pain of a dentist's drill faintly penetrating the anesthetic. Without adequate oxygen to metabolize his body's fuels, lactic acid has built up in his muscles to painful levels. Still the black gusts of current hold him down. He looks up through the green blackness toward the direction he thinks is the surface. His friends are up there waiting for him. They are reaching out for him. He can't disappoint them. They didn't want him to go down here. Now they want him to surface. He can't stay down.

He summons his whole concentration, like scooping up an armload of fallen leaves that want to waft away on the wind. *Up, up, up.* He strains up through the dark water to see the silvery surface. *Up is life. Down is death.*

More than actually seeing it, he senses the light again. He feels his body tossed upward. He hears swirling bubbles. White foam surrounds him. Cold air strikes his face. The roar of water fills his ears. Reflexively he exhales a great sigh of carbon dioxide. His mammalian instincts take over, and he pushes up his head like a seal pushing its nose through a hole in the ice. He breathes in. As he does, his body is tossed over into another hole. Matt's last breath is a breathful of foam.

0 minutes, 54 seconds (325 milliliters of oxygen remaining): Matt gags on the foam as he goes down again. His larynx is in spasms, reflexively closing on the water and foam. No more water can enter his lungs. Many drowning victims inhale only a glassful or so of water, and in "dry" drownings—about 10 to 15 percent of all victims—the larynx closes before inhaling and the lungs contain no water at all.

Matt doesn't know quite what's happening to him. He has a vague sense of tumbling, as if he's in a giant, warm whirlpool and helping hands are massaging and lifting him.

1 minute, 23 seconds (220 milliliters of oxygen remaining): Drowning doesn't happen all at once. It happens gradually. Matt's consciousness shrinks to a smaller and smaller spot, then blackens. He's tumbling along the dark and turbulent bottom. The water he inhaled has filled some of the alveoli sacs and "washed out" their surfactant, or protein coating. Without it, the lungs' inflatable sacs can stiffen and collapse, as in the case of premature infants who are

born with a surfactant deficiency and whose lungs are treated with substitutes made from cow or human amniotic fluid. Even if Matt were now brought to the surface and revived, he could perish a few hours later of "secondary drowning"—when the damaged lungs fill with fluid from the victim's own body, causing pulmonary edema and drowning the victim from within.

By now, however, there is little chance that someone will rescue Matt. Immersed in the crashing river sounds and brilliant daylight that bathes their ledge of rock, Matt's friends scan the river, straining to see where he's gone. They begin to worry. He should have popped up by now in a calm pool along the riverbank. They see nothing, not even his kayak. They don't realize that Matt—or what is becoming less and less Matt—is 8 feet below the river's surface and already a quarter mile downstream from the hole he tumbled into at House Rock.

2 minutes, 16 seconds (98 milliliters of oxygen remaining): Matt is like a rag doll. Concentrating his blood in heart and brain, his body makes its last stand against the approach of death. That last stand can be miraculously long in certain cases, especially among young children drowning in frigid water, when the diving reflex, for unknown reasons, responds profoundly and hypothermia slows the victim's metabolism nearly to a standstill, preserving a faint hint of life.

On February 6, 1974, at 11:30 A.M., in a classic and well-documented incident, a four-year-old Norwegian boy was seen to fall through river ice. By the time divers arrived from 15 miles away and located his body in 10 feet of water, the boy had been submerged for exactly forty minutes. His body was gray and lifeless, his pupils fixed and dilated, his pulse absent. Given immediate CPR, he was rushed to the hospital, arriving ten minutes later. Doctors ventilated his lungs with pure oxygen, rewarmed him on a water-filled mattress, and injected his blood with a solution of baking soda to correct its high acid content caused by so much carbon dioxide and metabolism without oxygen. The doctors kept up chest compressions for an hour after his arrival, when his heart kicked in strongly on its own. By that evening, the boy moved his eyes and stirred his limbs, and the next night he understood verbal commands. The day after that he was fully conscious. He left the hospital eight days after his admission in order to celebrate his fifth birthday at home. That spring—having fully recovered mentally and physically—he learned to ride a bicycle.

4 minutes, 21 seconds (37 milliliters of oxygen remaining): Matt is slowly ceasing to exist. He's not young enough and the water isn't cold enough for him to experience a diving reflex as profound as that of the Norwegian boy. His feeble heartbeat still pushes a residual amount of oxygen to his brain.

7 minutes, 55 seconds (trace amounts of oxygen remaining): Now alarmed, Matt's teammates scramble downriver along the rocky banks, searching for his kayak, his paddle, him. They see nothing. Matt's body is already half a mile down the rapids, tumbling at a depth of 12 feet.

19 minutes, 36 seconds: Matt's brain function has ceased but for the faintest flickers of electrical pulses. Two of Matt's friends begin to hike—run, really—up the steep, rocky trail out of Tiger's Leap Gorge to summon help. His other friends continue to comb the riverbank. Maybe he's dragging himself to shore somewhere downstream, out of sight. They hope against hope.

1 hour, 6 minutes: The searchers reconvene half a mile downstream from where they started on the ledge of rock. Even here, the whitewater is frightening in its violence, its tossing, foaming mounds of water shaped like haystacks and its sucking holes. Over the roar of the rapids, they shout back and forth to each other, waving their arms at various holes and pools of the river where Matt should be but isn't, their panic climbing, then subsiding like the ululating roar, until it's finally overcome by resignation and surrender to the obvious: Matt has drowned.

5 hours, 23 minutes: The two who went for help reach a small

Chinese village terraced into a mountainside. Inside the mud-brick government office is a single old black telephone. The Chinese official in charge picks it up to alert the authorities in the county seat. When he hangs up he looks at them for a few moments in silence, then speaks quietly in broken English. He tells them that the Chinese have no resources to spend on an elaborate search with helicopters or rescue divers, even if there were any around, which there are not. Chinese people fall into the rivers every day and drown, the official tells Matt's friends. You'll see them in the Yangtze, downstream, if you ride a passenger boat. Their bodies fill up with gases and rise to the surface and bob along like logs. No one has rescued them. Your friend has chosen to put himself in dangerous water. Now the dangerous water will carry him away to the sea.

Matt's friends rage against the official. They pound his desk with their fists and kick his office walls. But they know that their anger and frustration have nothing to do with the official or any attempts at rescue. Matt, they are sure, is long dead.

Finally they leave the office. They walk through the village back toward the gorge, cursing the official, cursing China, wishing they had never come. The mud-brick village with its red-tiled roofs and mountain backdrop that looked so charming when they first arrived now looks impoverished, callous, mean-spirited. At the far end of the village, they walk past the Taoist temple. They cannot read Chinese characters, but if they could they might note the inscription whitewashed onto its ochre mud wall:

> *The highest good is like water.*
> *Water . . . does not strive.*
> *It flows in places men reject and so is like the Tao.*

CHAPTER THREE

ARCHING TOWARD THE CLEAR LIGHT

MOUNTAIN SICKNESS

"Adrian's stumbling," Mara called out.

Beneath the rush of wind across the snow Mara heard her own breath—one, two, three, four times her chest heaved before she could take the next step up the slope. Just above, the three women of the first summit team waited. At her shout they turned their heads in unison down the slope. Mara could see the sharp flash of high-altitude sunlight off their dark glacier glasses as their eyes traced the 150-foot rope from Mara, who led the second team, down to Adrian, who was its middle climber.

Adrian's head was down. She lifted each foot excruciatingly slowly up the steps that the first team had kicked into the snow slope. It looked as if the sharp metal points of the crampons, strapped to the soles of her thick plastic climbing boots to prevent her from slipping, were adhering to the snow with half-cured epoxy resin. As the women watched, she inexplicably lurched to her left, dropped to one knee, and tried to catch herself from falling forward onto her face by leaning on her ice axe as if it were a cane, but it slipped from her grip and she sprawled in the snow.

She lay facedown, her arms making slow swimming motions as

if absently exploring a way to push herself up out of a pool of water. From the lower end of the rope, Linda, the last climber on Mara's team, shouted encouragement.

"Come on, Adrian, get up, you can do it! Just up to the others! Get up to the others and we'll rest!"

Slowly Adrian pushed herself to her knees, maneuvered one foot under her, then the other, and started up again.

The two were just an hour or so short of the summit. To get this far had taken them three weeks of hard labor, climbing without bottled oxygen and pioneering a difficult new route up the north face of the Himalayan giant—or rather giantess. This mountain was known to the natives as Annapurna, or "Goddess of the Harvest." At 26,504 feet, it was the world's tenth highest mountain, thrusting up out of the lush green rice fields in the center of Nepal in an enormous massif of a central peak surrounded by dozens of subsidiary peaks. Not as high as Everest nor as technically difficult as K2, the world's second highest peak at 28,251 feet in Pakistan's Karakoram Range, it was still a very formidable mountain. It was known for its beautiful surroundings and its remoteness, as well as for the enormous avalanches that swept down from its broad upper slopes and tumbled over the icefalls and glaciers thousands of feet to the valleys below. Its snow was abundant, its storms frequent and fierce, and its weather and avalanche risk could easily trap the unwary in places far higher than they wanted to be.

Mara reached the three women waiting above. "We've got to let Adrian rest," she gasped between breaths. "She's getting clumsier with every step."

"We can't rest too long or we're not going to make it," said Becca, the lead climber on the first-team rope. Becca was also the organizer of this all-women expedition, the one who had planned the summit attempt and kept everyone moving on schedule, delicately balancing their ambitions to make the top, their physical condition and need to acclimatize, and the fact that the monsoon was due to arrive any day, bringing with it storms that would prevent any further climbing.

Adrian finally kicked her way up the last steps to the little group. She sat down abruptly in the snow, head between her knees, back heaving.

"How are you feeling?" asked Becca. "Do you think you can keep going, or do you want to rest here for a few minutes?"

"I'm having a hard time today," was all Adrian said, her head still between her knees.

"What's going on?" Linda asked as she came up the last steps and completed the two teams gathered in a little clump on the vast white slope that rose to Annapurna's summit.

"We're resting," replied Becca, the way one might use the collective *we* when referring to a child.

Linda looked again at Adrian, head on knees, back and shoulders heaving for air. "Then what?"

"I don't know," said Becca. "Let's wait and see how she's feeling."

"Not up, I hope," Linda said. "Not when she's like this."

"Well," spoke up Gayle, one of the first-team climbers, "I'm not planning on going down, if that's what you mean."

The silk caravans of the ancient Chinese crossed two passes in the Karakoram Range named Little Headache and Big Headache, as recorded two thousand years ago by the Chinese official Too Kin. No one understood at the time that the earth's thinning atmosphere at high altitude triggered the symptoms of headache, nausea, fatigue, and a host of other problems that characterize mountain sickness. High-altitude travelers and residents offered many other explanations for the malady. The Tibetans, inhabitants of earth's highest plateau just to the east of the Karakoram, believed that travelers became sick when crossing high passes from inhaling the poisonous gases that swirled about the tallest peaks. On the opposite side of the globe, inhabitants of the Andes called the malady *mareo de punas*—seasickness of the high deserts—or *soroche*, a term also referring to antimony ore, as it was believed that noxious vapors from the ore sickened the laborers who worked the high-altitude mines. José de Acosta, a sixteenth-century Jesuit priest who traveled with the Spanish conquistadores over the Andes, wrote: "I was surprised with such pangs of straining and casting as I thought to cast up my heart too."

If you could weigh a column of air 1 inch square that reaches many miles up into the sky until the earth's atmosphere altogether vanishes, it would weigh at its base at sea level 14.7 pounds (thus the air pressure at sea level is 14.7 pounds per square inch). At 18,000 feet, however, it would weigh only about half of that, or a little more than 7 pounds, containing one-half the oxygen of the air at sea level. Nowhere in the world do humans live permanently above about 17,000 feet, the altitude of Peruvian mine sleeping quarters; above 19,000 feet the body can no longer acclimatize and slowly but progressively deteriorates, weakening and losing weight. An unacclimatized human suddenly placed without an oxygen mask atop the summit of 29,035-foot Mount Everest, where that 1-inch-square column of air weighs only about 5 pounds and contains one-third the oxygen, would lose consciousness in about a minute and a half and probably soon die. This is what befell the two companions of the balloonist Tissandier, who in 1875 made a rapid, three-hour ascent to 26,000 feet. He then passed out, and the balloon sank into an hour-long descent to the earth, whereupon Tissandier regained consciousness surrounded by his dead companions.

This is in stark contrast to birds, whose many adaptations allow them to tolerate extreme altitudes easily. Their lungs are something like a jet engine, in that they absorb oxygen continuously on both inhale and exhale, using a series of air chambers to maintain the continuous flow. The special hemoglobin in their blood snaps up oxygen even at very low pressures. In some birds, such as the bar-headed goose, the muscle tissue appears almost black due to the concentration of myoglobin, which helps store and deliver oxygen to the muscle cells' mitochondria, permitting the members of this species to follow their regular migratory route flying nonstop over the top of Mount Everest and the Himalaya between the plains of India and the highlands of Tibet. The highest-flying bird ever recorded was a Ruppell's griffon that whapped into an airliner while soaring on a warm updraft of air at 37,000 feet over the African plains. Like Hemingway's leopard lying in the snows of Kilimanjaro, no one could explain what it was doing up there. Even the common house sparrow pecking at the bird feeder in one's backyard can breathe the air at 30,000 feet, or 6 miles high, without any problem.

While it can't manage nearly as well as the sparrow, in its remarkable adaptability the human body is capable of adjusting to short stays at extreme altitudes (defined as those over 18,000 feet). A few powerful and well-acclimatized climbers such as Reinhold Messner and Peter Habler have even attained Everest's summit without breathing bottled oxygen, so close to the margins of the earth's thin atmosphere and of human endurance that for the last thousand feet of altitude they collapsed in the snow to rest every ten or fifteen steps. Over time—measured in days or even weeks—the body calls on a complex array of compensatory mechanisms to counter the lack of oxygen at high altitudes, adjusting its blood chemistry, fluid balance, and heart and breathing rates the way that a computer chip adjusts the fuel and air mix in the engine of a car to compensate for altitude. One study dramatically showed this adaptability to altitude by placing Peruvian highlanders in a special chamber and removing the air until the chamber's interior reached a simulated altitude of 30,000 feet. Eight of the sixteen highlanders remained conscious indefinitely, while the two sea-level dwellers, used as control subjects, blacked out in a minute and a half.

Like the computerized fuel-injection system, however, the acclimatization mechanisms in the human body sometimes go askew, especially when a visitor to high altitude doesn't take the time to adjust slowly and pushes too high too fast. The result is mountain sickness, whose symptoms are well known but whose exact workings are poorly understood. In its mildest form it consists of headache, tiredness, nausea, and a feeling of malaise. These symptoms are very common to visitors even at altitudes as low as 8,000 feet—the same as many Colorado ski resorts. Even a lowlander's visit to Denver, at only 5,000 feet, will cause some loss of night vision as the lack of oxygen affects the vision centers of the brain. In most cases, visitors to up to 8,000 feet become acclimatized after a day or two of rest, the headaches disappear, and they go on with their vacations. Only a very small percentage develop more severe forms of mountain sickness and must be hospitalized.

Rapid ascents above 8,000 feet boost the likelihood of both the less severe forms of mountain sickness as well as the two potentially fatal forms: high-altitude pulmonary edema (HAPE) and high-altitude cerebral edema (HACE). Roughly 3 to 5 percent of trekkers ascending to 14,000 feet in the Everest region were afflicted with mountain sickness and about 1 percent suffered pulmonary edema, according to one study. Pulmonary edema's incidence among climbers on Mount McKinley, where climbers ascend quickly with heavy loads in very cold temperatures, has been estimated at one in fifty. The low oxygen pressure and heavy exertion strain the lungs in such a way to fill their sacs with fluid, which, if left unchecked, drowns the victim from within. For many decades its symptoms were mistaken for those of pneumonia: extreme shortness of breath, fatigue, coughing, sometimes blood in the sputum. The symptoms of the less-frequent high-altitude cerebral edema, when the brain swells with fluid, imitate those of an evening spent at a bar: stumbling, lack of coordination, hearing or seeing things, drowsiness. The difference is that one may never wake from the coma that follows.

Treatment for both HAPE and HACE is really very simple and, if undertaken promptly, usually very effective. One high-altitude physician, with succinct urgency, describes the first three choices for treatment this way: "Descend! Descend! Descend!"

The problem is, it's not always possible to descend quickly. And in the places where one is most likely to suffer high-altitude pulmonary or cerebral edema—at 26,000 feet on a Himalayan mountainside—one is not likely to find a hospital.

"I don't think she's getting any better," Linda said after they'd been resting five minutes or so.

Becca placed her hand gently on Adrian's back. "Do you want a few more minutes?"

Without looking up, Adrian nodded.

"I think we should take her down right now," Linda said. "It might be exhaustion, or it might be something more serious. Either way, she can't go another hour like this."

Becca stood and looked up toward the summit ridge, silhouetted against the dark blue sky, a plume of snow spinning off it, and then the first-team climbers. No one spoke. Mara sensed the tension, the unspoken pressure on Becca, who had to make the decision. It would have been an easy decision if they weren't so close, if they knew they had more time, if the monsoon weren't due to arrive . . .

"Look, I'll stay with her if no one else will," Linda finally said. "Everyone else can go all the way up. Adrian and I will wait here until you get back."

"Are you sure you want to do that?" asked Becca. "This is probably going to be your only shot."

"She's my ropemate and my tentmate and my friend," Linda said. "I care more about her than I care about the summit."

Mara felt the group's attention shift to her, now that her two ropemates would stay behind.

"What about you, Mara?" Becca asked. "You can come with us, and we'll just rope together in two teams of two."

Mara tried to think it over. *Go or stay.* Her mind seemed to spin so slowly in the thin air, like a cloud gently tumbling over a peak. She'd climbed since high school, when she'd gone camping with a friend whose parents guided her up a small peak. She'd attended mountaineering camps during the following summers, eventually becoming an instructor, then went to college in a mountain town and studied environmental education with the hope that teaching about wild places could keep her close to the mountains. There were a thousand reasons she loved the mountains, and each mountain offered its own pull on her affections. Sometimes it was the sight of that pure white snow against the deep blue sky that compelled her, the sharp white edge between heaven and earth, or the tiny blue and pink wildflowers poking up so cheerfully among the weather-blasted barren rock, or the cold wind scrubbing her face and whipping tears from her eyes, or the satisfaction of solving a difficult problem—how to find a graceful and safe route to the top, cleverly using the ridges and couloirs and snowfields to help her toward her goal. Sometimes it was standing on top of a peak and feeling so tiny against the huge, indomitable mountain and the vast world spread out below. Sometimes it was the sense of teamwork when roped together and working as a single unit, and other times it was the independence and confidence it gave her. In a world that didn't easily recognize women's achievements and often tried to undermine them, climbing a peak was an undeniable and tangible attainment. No one could say she hadn't done it.

When Becca, whom she'd known along with Linda from their days instructing together at a mountaineering camp, told her she was organizing a major Himalayan climb for women, Mara jumped at the chance. All the compelling reasons she knew for climbing lesser peaks, she figured, would be magnified enormously by the sheer scale of the Himalaya. In addition, there was the nearness to Tibet. Annapurna's summit lay only about 30 miles—as the bar-headed goose flew—from the high ridge that marked the Tibetan border. She liked the notion that she could peer down from one of

the earth's highest peaks onto its highest plateau, one of the strongholds of Buddhism, a religion she'd been drawn to out of its respect
for living things.

She wanted to attain that summit in a manner she felt good
about. Leaving her two ropemates behind while she made the final
climb seemed an act that she'd end up regretting. Instead of grace, it
felt like selfishness. But maybe there was still a way she could catch a
view of Tibet without going all the way to the top.

"I'll stay," she finally said.

"Mara, you should feel free to go," Linda urged. "We can manage here without you."

"That's all right," Mara replied. "I'll stick with you two."

Adrian, head still down between her knees, raised her right
hand slightly to acknowledge Mara's decision, a gesture of thanks.

Adrian's problem had started several weeks earlier, on the two-
week trek from the low Nepalese valleys to Annapurna's base. With
hired porters hauling the heavy loads of tents, climbing gear, and
food, the women walked up the mountain trails, sleeping at somewhat higher camps each night in order to acclimatize slowly as they
ascended from the rice fields up to the hill farms. By the time they
reached the higher villages at about 10,000 feet, their bodies had begun to respond to the thinning air—but, at first, in two curiously
contradictory ways. Two sensors in their bodies, each adjusting to
the new altitude, were working against each other. As they climbed
into the thinning air, the oxygen sensors in their carotid arteries noted
the dropping level of oxygen in their bloodstreams and sent out messages telling their lungs to suck in more air with each breath. But the
more deeply the lungs breathed in, the more carbon dioxide they expelled with each outbreath. The carbon dioxide sensors residing in

the brain noted the falling carbon dioxide levels in the women's bloodstream. They told the lungs to slow down—the women were altering their blood chemistry by breathing so hard.

They needed to acclimatize. Slowly they made their way beyond the last of the trees to Base Camp at 15,000 feet, where they pitched their tents among the heaps of broken, tumbled rock, known as moraines, pushed down the mountain by glaciers. After nearly a week at high altitude, their bodies had engineered a peace pact between the warring mechanisms. Their kidneys, serving as mediator, cleverly tricked their carbon dioxide sensors into thinking their blood contained more carbon dioxide than it actually did by removing some of the blood's alkaline substances, thus making the blood more acidic, as if it contained carbon dioxide that needed to be expelled. Now, instead of contradictory messages, both the oxygen and the carbon dioxide sensors were sending the same urgent signal to the lungs: *Breathe, and deeply!*

This physiological urge to breathe deeply at high altitude is a good thing—an adaptive mechanism—but varies widely among individuals. A strong hypoxic ventilatory response (HVR), as this urge is known, helps one adjust more easily to altitude, while a weak one, it is suspected, makes one more susceptible to mountain sickness. Contrary to what one might expect, there is evidence that training for long-distance sports may actually weaken rather than strengthen one's HVR. It has been proven over and over again that climbers who are in top physical shape are just as prone to mountain sickness as those who are not. Perhaps also contrary to expectations, men have a weaker HVR than women; female hormones such as progesterone and estrogen help stimulate breathing. In one study conducted atop Colorado's 14,000-foot Pike's Peak, the extra oxygen in women's blood, probably due to this increased breathing, was the equivalent of knocking 540 to 1,640 feet off the summit for the women in the group compared to the men. Researchers speculate that this may be why women have a substantially lower incidence of high-altitude pulmonary edema than men.

But they are not immune to it.

As the group trekked to Base Camp, Adrian had suffered head-aches and nausea. After they'd spent a few days quietly in camp, these had subsided, aided by twice-daily doses of acetazolamide, or Diamox, a drug that increases the respiratory rate by reducing the alkaline content of the blood. The team then slowly began to estab-lish higher camps up Annapurna's north face, ferrying the loads from one camp up to the next. They'd been careful to observe a gen-eral rule of thumb, not sleeping more than 1,000 to 2,000 feet higher each night and taking frequent days off to rest. It is the sleeping altitude and not the maximum height one reaches during a day's climb that largely determines how profoundly one is affected by alti-tude. Thus the climbers' adage, "Carry high, sleep low."

Once on the mountain Adrian felt better. From Base Camp, they climbed the glacier at the mountain's foot, marking a winding route through the crevasses with wands, and established Camp I. Above it came some of the most difficult climbing. They ascended nearly sheer ice faces on the mountain's lower ramparts, kicking the spiky front points of their crampons into the ice and planting the picks of their two ice axes with solid thunks so they were clinging to the faces like spiders on a windowpane. Finally, they screwed into the ice threaded alloy tubes—ice screws—and to these the women fastened permanent ropes for safety and assistance as they ferried loads up the walls to establish the higher camps on the less-steep slopes above, like rungs on a ladder.

By the time the women established Camp V, just under 25,000 feet, their physiologies were undergoing other odd transformations. Their bodies were busily producing extra red blood cells to carry more oxygen. Their heart rates when lying in their tents rose from a normal pulse of about 50 to about 90 beats per minute, and their maximum heart rates when carrying heavy loads dropped from about 180 per minute down to 140, until, if they'd gone as high as the summit of Everest, their resting and maximum rates would nearly equalize, as one climber's did, around 115. At these altitudes, there was simply no longer enough oxygen to keep the heart and all the organs fully supplied, and more and more of the oxygen was siphoned off by the hard-working muscles that pumped their panting lungs. By Camp V, their climbing rate had dropped way down, to about 500 vertical feet per hour, compared to over 2,000 feet an hour at sea level. (Near the summit of Everest at 29,000 feet, the rate would probably bog to 100 or fewer feet per hour. Even the great Messner, who near sea level could climb in one hour the equivalent of five World Trade Centers stacked on top of each other, or more than a vertical mile, took an hour to drag himself, without supplemental oxygen, up Everest's last 300 feet.) Had they been Sherpas, Tibetans, or Peruvian highlanders, the womens' bodies may also have compensated for the lack of oxygen in still more subtle ways, down inside the mitochondrial furnaces of the cells, where high-altitude rabbits and other mountain animals appear to metabolize their bodies' fuels at a faster rate than their lowland counterparts.

After three weeks and their long, slow acclimatization, everything was in place for the summit attempt. As the weather cleared to perfection and Becca gave the go-ahead for the summit, the first team happened to be up at Camp IV and the second team down at Camp III. This meant that for the second team to catch up while the good weather lasted, they had to climb long and hard in one day all the way from Camp III to Camp V, the jumping-off point, gaining over 4,000 vertical feet in one day and thereby violating their rule of thumb.

When she woke on summit day at Camp V, Adrian complained of feeling nauseous and tired but said she'd try for the summit anyway. They'd climbed through the pink dawn without problem, ascending above the sun-touched, ice-capped lesser peaks that surrounded Annapurna, until about midmorning, when Linda and Mara noticed her stumbling and weakening. Finally they'd called out to the first team to stop.

"Are you ready to try descending?" asked Mara.

Adrian, still sitting in the snow, nodded, as if too tired even to speak, and clumsily moved her boots beneath her in an attempt to stand.

Mara and Linda each took an arm and helped her to her feet. While Mara belayed them from above, Linda wrapped her arm around Adrian's waist and propped her up as they descended. They'd been climbing together for years, and Adrian was the godmother to Linda's daughter. But even with Linda's help, Adrian frequently stumbled. Sometimes Mara would have to climb down to help.

Once, as they were pulling her back up, Adrian suddenly turned to Mara. "I know what you two are trying to do. You're trying to cut me out of this climb."

Mara looked closely to see if Adrian was joking, but her jaw was thrust forward, and behind the dark glacier goggles her eyes appeared wide with anger. Irrational behavior was one symptom of high-altitude cerebral edema.

"Adrian, you're not well," said Mara. "We're taking you down so you'll be better."

"You want the summit for yourselves," Adrian countered. "I'm on to your scheme."

"There is no scheme, Adrian."

"I thought you were my friends, and now you're betraying me."

"Adrian, we're trying to help you," Linda said soothingly.

"I don't want any help from you backstabbers," Adrian fired back, and sat down abruptly in the snow.

Mara turned to Linda, speaking quietly over Adrian's head. "Let me try something." She then bent down and spoke clearly. "Adrian, I'm going to walk to that knoll over there. I'll be right back."

Adrian said nothing and slumped further in the snow, as if now too fatigued to respond.

Mara untied from the rope and set off alone, traversing across the slope. She hoped that from the knoll she could finally see down into Tibet, and that her brief absence might jog Adrian from her angry paranoia. Working carefully, planting the pointed shaft of her ice axe into the snow to serve as a self-belay to hold her if she slipped, Mara steadily made her way across the slope. But the knoll lay farther off than it looked, and it took Mara nearly half an hour to reach it. Kicking a few steps into the snow to its small summit, she realized the view of Tibet was still blocked by a section of ridge line. All she could see were a few black specks that might have been birds winging past. Looking down from the knoll, Mara also noticed the first misty clouds of what looked like an oncoming storm, billowing upward from the valleys below like a cottony quilt. The sky still shone blue above.

With a rising sense of worry, Mara hurried back across the snow slope, or hurried insofar as she could, hyperventilating at just under 26,000 feet. They'd have to move quickly to descend to Camp V before the clouds rose and engulfed the upper mountain. She looked up, hoping to spot the first team. High above on the snow slope, against the blue sky, three black dots were making their way down, tiny orderly points on the vast luminous ceiling of white and blue. They must have made the summit.

The wind was rising when she reached Adrian and Linda.

"She's been talking to you while you were gone," Linda said, speaking just loudly enough to be heard over the hiss of snow blowing at boot level across the slope. "She thought you were sitting right beside her, even when I pointed you out over there."

Known as the "phantom companion," this hallucination is not uncommon among those suffering mountain sickness, especially cerebral edema, at extreme altitudes. Climbers have spotted highway crews working at 20,000 feet on an Andean peak when the nearest road was many days away, and offered pieces of candy to climbers who weren't really there beside them on Everest. One recovering victim of high-altitude cerebral edema kept seeing Marilyn Monroe moving about in his hospital room. In one epic incident in 1981, a Swiss climber lingered a few extra minutes late in the afternoon after his friends had started down from the 23,191-foot summit of

Glacier Dome, just east of Annapurna. He soon found himself accompanied by a German-speaking man who pointed out a shortcut down to the party's highest camp. Following the advice of the stranger, who soon disappeared, and descending into the darkness, the climber spent a forced night in an icefall at temperatures of around −40 degrees Fahrenheit. He suffered severe frostbite and nearly died in an avalanche, hallucinating at one point that his friends had constructed a cable car up to him but wouldn't let him board it. They did finally show up—on foot and not in a cable car—and rescued him.

Mara and Linda pulled Adrian, now nearly limp, to her feet and slowly moved down the mountain again, taking turns propping her and at times belaying each other. Adrian's feet bumped into each other and her crampon points caught in the fabric of her gaiters, the protective sheath that covered her climbing boots and boot tops to keep the snow out, tripping her, so that she stumbled drunkenly until Mara or Linda dragged her upright again. It was excruciatingly tiring. Mara felt she was carrying a sack of cement through deep snow as she and Linda took turns struggling under the additional weight. Adrian was of no help at all. She didn't seem to care if she went up or down, if her mittens were on or off, if she was helped or not. Mara focused on one thing: getting them all safely down to Camp V.

The stumbling and lack of coordination—known as ataxia—the lassitude, and the inability to care for oneself all are classic symptoms of high-altitude cerebral edema. No one knows exactly why it would have afflicted only Adrian and not the others. Some researchers believe that everyone's brain swells during rapid ascent to high altitude but some individuals tolerate the swelling better than others. The precise cause of the swelling is also a mystery. Blood flow to the brain jumps as much as 25 percent when someone—anyone—first ascends to high altitude, but as acclimatization occurs over several days it drops back down. This initial high flow, however, might rupture brain capillaries and leak fluid and blood into brain tissues, causing the swelling. Or perhaps the leaks in the capillaries are microscopic holes opened by the sprouting of tiny new branch capillaries—known as angiogenesis—which is one of the many ways the body attempts to deliver more blood to oxygen-starved tissues.

In either case, the brain swells. Floating in its bath of cerebrospinal fluid, the brain is totally enclosed in the tight space of the skull. As it swells, something has to give. If not, the pressure inside the skull builds enormously. There is, however, a kind of overflow valve for the fluid, and this valve may work better in some people than in others. As the brain expands, it squeezes excess cerebrospinal fluid from the skull down into the hollow centers of the uppermost vertebrae. Depending on individual physiology, some people have more space in their vertebrae for the excess fluid and more space inside their skulls to handle the swelling, perhaps making them less susceptible to high-altitude cerebral edema. Perhaps Adrian was the only one who suffered because she had an unusually tight fit between her brain and skull and little space in her upper vertebrae.

By the time the first team approached Mara and Linda and Adrian slowly making their way down the slope, windblown wisps of cloud, luminous at first with sunlight like brilliant shards of fog, were whipping past the mountain and coloring the deep blue of the sky with shades of pastel and sending patches of sun and shadow scudding across the broad white face of Annapurna's upper slopes.

"You made it?" Mara called out as they neared.

Becca raised her hand in a quiet gesture of success. Mara noticed that she and the other first-team climbers said no more about their victory, as if self-conscious about the price at which it might have come. Instead, the first words Becca spoke when the two groups finally joined on the slope were "How is she?"

"Very bad," Linda replied.

Everyone knew there was no time to be lost. The weather was shifting fast. The first team quickly moved ahead to break trail through the snow on gentler slopes where they didn't need belays, while Mara and Linda walked on either side of Adrian, propping her up. The ragged shards of flying cloud thickened into a featureless white mist whipped by the wind until Mara could barely see the members of the first team ahead of her. With the white clouds and white snow, she could no longer tell which was upslope and which downslope. It made her head feel like it was spinning. Spots danced before her eyes. It soon became obvious that they weren't going to make it down to Camp V that night. Adrian was falling continually, the other women were exhausted, daylight was dimming, and they were having trouble staying on route. They'd carried neither tents nor sleeping bags that morning in order to be as light as possible for the summit attempt.

"I think we should dig a snow cave while we still have some light," said Becca at a halt.

Exhausted as they were, no one spoke against it. They kicked back and forth across the slope with their boots to find the proper snow in which to build a cave. They needed a deep, windblown drift of snow that had compacted into a firm consistency, so the cave would hold its shape, but still soft enough so they could dig easily in it. Finally, stumbling through the whiteout to the left of their route down, they plowed into a deep drift. As the light dimmed and clouds swirled, they took turns with the two lightweight shovels they packed and bored a tunnel sloping upward into the drift, scooping the snow behind them and shoving it out the tunnel, laboriously hollowing out a cavern. The tunnel had to slope upward so that the warm air from their bodies, which rises, would be trapped in a kind of bubble inside the drift. Inside the cavern, they left a raised platform of snow across the rear half of the cave on which to sit, as the Inuit did in their igloos, so they'd be elevated above any cold air that pooled on the floor. Poking upward with the shaft of an ice axe, they bored a small air vent through the roof, and fashioned a block of snow to seal the doorway. Snow, consisting of mostly air, has very formidable insulating properties. With subzero temperatures prevailing outside, Inuit families used to lounge topless inside the comfort of their snow igloos in the light and warmth of their seal-oil lamps.

Pulling Adrian inside, the women propped her on the bench, first laying down empty packs to insulate her from the snow, and pushed the snow block in place over the tunnel entrance. Becca produced a candle from her emergency kit. The cave took on a cheery yellow glow from the candle light reflecting from the sparkling walls, and slowly began to warm up.

Situating themselves in the candlelit cavern, the climbers drank from their water bottles and ate a few pieces of candy. Wedged on the snow bench between Linda and Mara, Adrian appeared a little livelier. How was she feeling? Okay. Did she want water? Yes. It appeared that even the drop in altitude of only 1,200 feet or so had helped relieve her symptoms. Plus, at Becca's urging, Adrian had tried hyperventilating, which infuses oxygen into tissues and brain and can help relieve symptoms. Still, if they asked her to subtract one number from the next or if she knew where she was, she'd only shake her head. Mara wondered if Adrian knew she was in a snow cave at 25,000 feet on a Himalayan mountainside.

Becca took out a small stove that she'd packed for just such a situation, plus a little pot, and heated some instant soup that the women passed among themselves. Mara could feel the warmth trickling into her. In early evening at their regular appointed time, Becca moved to the tunnel entrance and called by handheld radio down to Base Camp to explain the situation. Base Camp radioed back medical advice. From the high-altitude medical kit that Becca carried, they should give Adrian an injection of dexamethasone—a steroid that acts as an anti-inflammatory agent and has been shown helpful in treating mountain sickness and cerebral edema—and give her tablets of it every six hours thereafter. Base Camp also told them to try to keep her in a sitting position.

Using her headlamp, Becca carefully gave the injection in Adrian's buttock. Everyone now knew there was nothing to be done except wait for first light.

After about an hour in the cave, eating and drinking and situating themselves, Gayle, of the first team, spoke up.

"We should blow out the candle."

"Why should we blow out the candle?" Linda replied testily. "The light's doing Adrian good."

"Because we might need it later."

"I think we need it more now," Linda said.

"Who knows what the weather will be tomorrow," Gayle replied evenly but coldly.

"Let's not think about tomorrow's weather," Linda shot back. "Let's think about now."

"Okay," Becca interjected. "Let's burn the candle another fifteen minutes or so and then blow it out and try to sleep. We can still use our headlamps."

With the candle out, utter blackness filled the cave. There was no sound of the windstorm outside, its scream muffled by walls of snow several feet thick. Mara heard only the small scrapes and crinkles of the cold-stiffened fabric of their climbing suits as the women made small adjustments to their positions on the bench, and beside her she heard the quick panting of Adrian's breath.

Adrian's breathing seemed to grow faster the longer they stayed in the cave, instead of slowing down with rest, as Mara's own breathing had, and was punctuated by fits of coughing. This fast breathing—or dyspnea—while at rest is one of the warning signs of HAPE, or high-altitude pulmonary edema, which often accompanies and may in some cases trigger high-altitude cerebral edema. Like HACE, no one knows exactly why or how HAPE occurs. It appears that the lack of oxygen at high altitude causes the blood vessels to constrict in certain parts of the lungs but not in others, at least in those who are susceptible to the illness. Large volumes of blood are forced through the vessels that remain open, the theory goes, forcing open tiny holes in their walls and spilling fluid and red blood cells out into the lungs' air sacs and passages. With a stethoscope, or even by placing her ear under Adrian's clothes and against the bare skin of her back, Mara would have heard a rasp or bubbling sound with each of Adrian's breaths, a condition known as rales or crackles, as the fluid, slowly beginning to suffocate her, moved in her air passages.

Mara feared that what she was hearing might be a case of pulmonary edema in the making. She was sure Becca suspected it also, and perhaps the other women, too. No one said anything, not wishing to alarm Adrian and thinking there was nothing they could do until morning anyway. Mara tried to ignore the panting breaths of Adrian beside her and the crinklings of the fabric, tried to ignore her own uncomfortably slouched position against the back wall, the crunch of the snow against her head. She tried to lie down, but the bench was too narrow and her legs projected over the floor. She shifted back to slouching.

Sleep seemed impossible. She tried to meditate, following her daily practice at home, when she would sit cross-legged on the soft, carpeted floor of her hushed bedroom, but here it was so much harder, with the crinklings and rustlings and pantings, with the cold scrape of snow on her back, with the raspy *hacckk . . . hacckk . . . hacckk . . .* of Adrian's cough. It was like trying to meditate inside a darkened refrigerator full of breathing, rustling living things. Mara gave it up. Instead, she attempted to sleep, trying to snuggle into the snow wall at her back. All around her, she could hear tiny, impatient noises.

"Okay, who brought the chips?" someone finally said in the darkness.

"And the Diet Pepsi," said another voice.

"And the beer and the music," said a third.

They started to laugh, the giddy, black-humored, semihysterical, tension-breaking laughter of people in an impossibly difficult place, abetted by the scant oxygen making its way to their brain tissues.

"Don't forget the men," someone said.

"Don't you remember? We left them three vertical miles below us."

"They're down there doing the housework while we're up here enjoying ourselves."

More giddy laughter. Even Adrian seemed to be laughing; Mara felt the rhythmic jostling against her shoulder, although it wasn't clear if Adrian caught the jokes, her clarity of mind sharpened somewhat by the injection of dexamethasone, or whether, in her frail, oxygen-starved state, she was simply carried along by the others.

And so it went through the night in the snow cave. After the giddiness subsided, they told stories about their first climbs, their first loves, their favorite music, their idea of what, at this very moment, would make a perfect dinner menu, whether dessert should be hot apple pie, or cake, or a stack of pancakes dripping maple syrup.

They periodically dozed, and when someone was asleep Mara could hear the peculiar disturbed breathing of extreme-altitude sleep, when climbers inadvertently wake themselves up to 150 times per hour as the body falls victim to its own vicious respiratory cycle. The lack of oxygen in the blood triggers the oxygen sensor, which tells the lungs to hyperventilate. The hyperventilating expels carbon dioxide from the blood, triggering the carbon dioxide sensor, which tell the lungs that there's no hurry to take another breath. The lungs stop breathing. The oxygen level in the blood suddenly plummets, waking the sleeper with the sense that he or she is suffocating, and triggering another bout of hyperventilating and the cycle begins again, over and over. Even acclimatization can't break the periodic-breathing cycle that occurs at extreme altitudes, and some researchers believe that the reason humans can't live permanently above 19,000 feet is that they get so little sleep, their bodies simply fall apart.

Though she slept fitfully, Adrian's cough worsened throughout the night. Toward dawn, she fell into long hacking bouts that brought up fluid from her lungs; bent over at the waist, she let the fluid fall from her mouth into the snow between her feet. When Becca momentarily shone her headlamp on the thin, pinkish sputum, bloody streaks were visible.

"This is not a good sign," she said. "Not a good sign at all."

Becca took her pulse, timing it with her wristwatch, and calculated it at about 130, her respirations about 35 per minute compared to a normal 12 or so at sea level. Both heart and lungs were working furiously to bring oxygen to her tissues, but the oxygen from the air simply wasn't being absorbed by her saturated lungs. Now they could all hear in the periods of silence the gurgling sound inside her lungs with each inhalation and exhalation, and imagine in the darkness the foam around her mouth.

Her condition was now progressing to severe high-altitude pulmonary edema. Meanwhile, her brain was still swollen with cerebral edema.

They decided to wait until 6 A.M., when they'd have enough light to descend, to unplug the snow block from the cave's entrance. If things went well and Adrian could walk, they could reach Camp III by afternoon, where at 21,000 feet the air was noticeably thicker and could help her greatly. A descent of even only 1,000 or 2,000 feet can bring about substantial improvement from pulmonary edema; the early Spanish accounts from Peru used to describe this as recovery *por encanto*—as if by magic. The women stowed their bits of gear in their packs, tightened the straps of their crampons, zipped shut their clothing. Every few minutes they checked their watches. At exactly 6 A.M. Gayle crawled down the tunnel and pushed aside the block of snow. Mara saw the dim gray light filter into the tunnel, followed by a cold rush of air, then the muffled shriek of the wind.

"What's it like out there, Gayle?" Becca called out.

"It's a whiteout," she called back. "It's a frigging blizzard."

"Can we descend?" Becca said.

Gayle pushed herself farther out the tunnel, then slid back in.

"Not a chance. We'd be totally lost within the first hundred feet."

She shoved the block of snow back into position, casting them into darkness again, and crawled backward into the cave, the beam of her headlamp glancing about as she maneuvered back onto the bench.

"I hope this isn't the start of the monsoon," she said.

"If you hadn't insisted on summiting," Linda said out of the darkness, "we wouldn't be sitting here worrying about monsoons. We'd have Adrian down at Camp Three by now, and she'd be recovering, and we'd all be in our sleeping bags drinking hot tea."

Gayle's headbeam swung through the darkness to fall on Linda.

"Didn't you want to summit?" Gayle said, her presence in the darkened cave only an angry voice and a cloud of vapor rising into the beam. "If you didn't want to summit, what did you happen to be doing hurrying up to twenty-six thousand feet on Annapurna? Looking for a decent café?"

"I wanted to summit," Linda said, switching on her own headbeam. "But I also know when to back off. No summit to me is worth putting someone's life in serious jeopardy."

"Don't go blaming me for where you find yourself," countered Jill, the other member of the first team, her voice rising. "You can claim you're climbing for the beauty or the camaraderie or whatever, but every time you set foot on a mountain, you're putting your own life in danger, plus the life of whoever's at the other end of the rope. If you didn't want the risk, you should have stayed home and watched TV."

"All right! Enough!" shouted Becca. "Instead of arguing, why don't we figure out what we're going to do."

But there really wasn't much to do except wait for the weather to clear. Sitting there, Mara felt her own sense of guilt rising. If she hadn't been so eager to traverse over to the knoll and see Tibet, she and Linda might have had enough time to bring Adrian down to Camp V before the storm hit. And now here they were.

Becca, always thinking of logistics, moved down the entrance tunnel, opened it partway, and made another radio call to Base Camp, telling them the weather was too socked in to move. One of the climbers in Base Camp had heard of a team trapped by weather relieving pulmonary edema by physically squeezing the fluid out of their teammate's lungs. Linda and Mara, following the instructions conveyed by Becca, took turns crouching behind Adrian, who was bending over with her head between her knees, wrapped their arms around her torso, and squeezed hard whenever she coughed, forcing the fluid from her lungs.

It gave some relief for a while, but then they'd have to do it again. And again.

Water was now a serious issue, for they'd long ago drained their bottles. Although their sense of thirst was blunted by the extreme altitude, their bodies also needed far more water—4 to 6 liters per day, compared to less than 3 when resting at sea level. Once above 22,000 feet they, like all climbers, inevitably became dehydrated. The cold, dry air at extreme altitude evaporated the moisture from their throat and lungs at a ferocious rate. It was difficult, up where there was no free-running water but only snow and ice, to melt enough each day to drink. Stoves didn't work efficiently. Water boiled at such a low temperature in the thin air that the ice was slow to melt, in water that might be boiling but not very hot.

Mara helped Becca set up the little stove, carefully balancing it on the packed snow of the floor and igniting it. Each act demanded long and careful forethought, as if the mind couldn't sprint but only place one foot ploddingly in front of the other. Memory and conceptual reasoning both drop off sharply above 14,000 feet. Cognitive performance tasks—the equivalent of assembling some mechanical device—could still be performed at extreme altitudes but at a considerably slower pace. Despite these short-term mental deficiencies brought on by hypoxia, there has never been any conclusive scientific proof—contrary to what many climbers believe—that long stays at extreme altitudes cause permanent intellectual impairment. Take the case of eminent Danish physicist Niels Bohr. As he was spirited from Sweden to England on a high-altitude flight to escape the Nazis during World War II, Bohr didn't understand the pilot's instructions to turn on his oxygen. Bohr endured altitudes up to an estimated 30,000 feet and several hours above 15,000 feet. Lifeless when he landed, Bohr recovered fully and went on to a distinguished career in atomic research.

The women scraped bits of snow and ice from the walls of the cave and laboriously fed them, a few chips at a time, into the little pot. Slowly they managed to fill three empty bottles. A fourth they'd already employed for a "pee bottle," used when it's too difficult to relieve oneself outside the tent, which they emptied in a corner of the entrance tunnel.

They sat silently much of the time, trying to doze, turning inward to their thoughts. Adrian's cough worsened as the day went on, becoming wetter, and her panting grew harder as fluid slowly filled the sacs and passages of her lungs.

"I can't get enough air," she said, one of the few times she spoke all day. "It feels like someone's sitting on my chest."

These few words sent her into spasms of hyperventilation and coughing. If they were standing outside the cave in daylight, the women would have noticed the bluish tinge—known as cyanosis—that the lack of oxygen had given to her fingernails and lips, to her nose, tongue, and ears. Her symptoms would begin to clear in just a few minutes and she'd recover completely in twenty-four hours if she were breathing high-flow oxygen through a face mask—6 to 12 liters per minute, compared to the much lower flows of 2 to 3 liters per minute that a climber attempting a summit would use. But carrying enough oxygen to treat a pulmonary edema victim would be extremely burdensome. A lightweight alternative, weighing about

15 pounds when folded up, is the hyperbaric bag. The victim is zipped into this large, airtight fabric sack like some giant frozen burrito sealed inside its plastic pouch while the victim's teammates, working a foot pump outside, inflate and pressurize the bag, giving the effect of 5,000 to 6,000 feet of descent.

But the women had neither supplemental oxygen nor a hyperbaric bag. Descent was their only option. Every hour, one of the women crawled down the entrance tunnel, shoved the snow block open a crack, and reported: "The same."

By midafternoon, Adrian could no longer sit upright. She slumped over first on Linda and then on Mara. Semiconscious, she seemed to be drifting in and out of sleep and couldn't coherently answer their questions. It was difficult for them to go on performing the squeezing maneuver, and it was now clear that, with the storm still raging, they would have to spend another night in the snow cave. Helpless to do anything else to assist her, frustrated at their helplessness, the five women went to work to enlarge the snow cave so Adrian could rest more comfortably. Taking turns with ice axes and shovels, they hollowed out a side chamber, shoving the excavated snow down the entrance tunnel. When it was Mara's turn with an ice axe, she kneeled in the side chamber with her headbeam bouncing over its back wall as she chipped away at belly level. Suddenly her axe broke through the wall. Gray daylight and a rush of snow swirled into the chamber. She carefully leaned forward and peered out.

"Oh my God!" said Mara.

"What do you see?" asked Becca.

Mara was kneeling on the brink of an enormous cliff of rock and ice that dropped away and disappeared into the swirling mist. Without realizing it in the previous afternoon's whiteout, they had dug their snow cave into a cornice—the sculpted wave of windblown snow that forms at the top of cliffs. During the night in their snug snow cave, their backs had rested on the other side of the snow wall on the edge of a thousand-foot drop.

Mara plugged the hole with a block of icy, semitransparent snow that kept out the wind but allowed a faint light to enter the cavern. On the bench they'd carved for Adrian on the side opposite the window, the women carefully laid down an insulating layer of backpacks and carried her into the side chamber.

But by evening, Adrian had slid into a coma.

Outside, the wind still raged, but they no longer bothered to check. The women took turns in the side chamber watching over Adrian. They were very quiet. Becca's stove was almost out of fuel. Their bits of food were mostly gone. If the storm didn't relent by to-morrow, their bodies, badly dehydrated and weak from a lack of food and the extreme altitude, wouldn't have the resources to ward off the cold. Nor would their legs have the strength and balance to carry them down the mountain. The saliva had dried on the roof of Mara's mouth into a thick coating. She could feel the cold seeping up through her feet and hands like a palpable force, stiffening her fingers and toes, sending sudden chills through her body in twitches and shivers, reminding her of the dissected frog's legs that twitched when a certain nerve was touched by a scalpel.

No one mentioned it directly, but their own survival now had come into question.

"This storm has to let up before too much longer," was all that they said to each other.

Sitting through the long hours, each woman reacted in her own way to the growing desperation of their situation, as if their own mortality and their fears about it were mirrored in Adrian. Linda spent long stretches kneeling beside her old friend massaging her legs and arms, whispering to her, "Hang on, hang on," telling her to keep fighting, that the weather would clear and they'd take her down soon. But then Linda would start sobbing—sobbing for the loss of her friend, sobbing, perhaps, in fear of her own death. Becca, at least outwardly, kept up a stoic front. When it was her turn, Becca sat beside Adrian and constantly fussed with Adrian's clothing and insulation, trying to make her more comfortable, as if simple efficiency and a sense of order—putting everything in its exact proper place—could restore Adrian to consciousness. Gayle, in contrast, refused even to enter the side chamber, huddled alone along the farthest wall of the main chamber, turning inward to escape the inescapable end.

"Couldn't you at least spend a few minutes with her?" Linda finally said. "You're supposed to be her teammate, after all."

"I'd really rather stay here," Gayle replied.

Mara, in her turn, held Adrian's mittened hand and tried to remember what she'd read from *The Tibetan Book of the Dead*. It is the belief of Tibetan Buddhists that the art of dying is as important as the art of living. While Westerners tend to fear death, Tibetan Buddhists welcome it as an opportunity to break free of the endless cycle of rebirth and suffering. But the mind of the dying person must be trained properly to seize the moment of death and ride it to liberation. To this end, *The Tibetan Book of the Dead*, as it's known in English—or to Tibetans, *The Great Book of Natural Liberation Through Understanding in the Between*—is read aloud by a lama into the ear of the dying or dead person. It serves as a kind of guidebook or road map through the labyrinths that one negotiates between death and rebirth—the period they call the Between—and includes chapters such as "The Out-of-Body Reality Clear Light," "Encountering the Lord of Death," and "Choosing a Good Womb."[1]

One practice Mara had studied that could be used to help a dying person is called Tonglen, or giving and receiving. It is an exercise in compassion—the greatest gift, the Tibetan Buddhists believe, that you can give to a dying person. Sitting in the darkness on the snow bench, Mara imagined all of Adrian's suffering as a hot, black, grimy cloud of smoke. She inhaled that cloud of black smoke on her in-breath. She took the hot, black cloud into her own heart and

---

1. *The Tibetan Book of the Dead: Liberation Through Understanding in the Between*, translated by Robert A. F. Thurman (New York: Bantam Books, 1994). For the practice of Tonglen meditation, see *The Tibetan Book of Living and Dying*, by Sogyal Rinpoche (New York: HarperCollins Publishers, 1993).

drew it into the worst part of herself, the grasping, selfish part. Mara used that hot black cloud to destroy the part of her own being that was selfish and grasping, purifying her own heart. Then she breathed out, exhaling to Adrian a cool, white, soothing light of peace and joy, compassion and love, that purified Adrian's negative karma, too. Even if Adrian wasn't conscious, her karma benefited from the practice, as did Mara's, the Tibetans believe. "For me," writes one Tibetan lama, "every dying person is a teacher, giving all those who help them a chance to transform themselves through developing their compassion."

Adrian's own breathing now had grown shallower. Becca lit the remaining stub of the candle, bathing the walls of the snow chamber with its warm yellow glow. Spontaneously, the other women crowded into Adrian's chamber, one after the other. Even Gayle squatted in the doorway. To Mara, it felt as if Adrian were surrounded by midwives who, instead of helping her give birth, were trying to ease her passage into death, back into the time before life.

The Tibetan Buddhists believe each dying person "dissolves" in the same manner, even someone undergoing a violent death, although the stages may pass extremely quickly and not always in the same order. First the outer senses dissolve, and the dying person cannot recognize the acquaintances at her bedside. Then the four elements withdraw from the body. The dying person cannot sit upright after the earth element withdraws. The bodily fluids leak out when the water element withdraws. As the fire element withdraws, the person's warmth seeps away. Finally, the air element withdraws. The dying person pants and rasps, hallucinates, and has visions—frightening visions if he or she has led a negative life, and pleasant visions of old friends or heavenly places after a life of kindness and compassion. The in-breaths become shallower, the out-breaths longer, and the dying person gives three long out-breaths. Breathing suddenly stops, and all vital signs vanish. At this point, the inner dissolution has just begun.

Just before 3 A.M., with Linda hugging her on the snow bench, Adrian exhaled her last rasping breaths. Linda remained there, hugging her. The others sat and squatted and knelt silently in the chamber while the steady yellow flame of the candle burned near her head.

After a few minutes, Linda moved away. Becca came forward and removed Adrian's heavy mitten and felt for a pulse.

"She's gone."

They sat there for a long time, each in her own thoughts, while the candle slowly burned down.

In Western medicine, the generally accepted legal time of death is when the entire brain, including the brain stem, which controls breathing and other basic processes, ceases to function. In Scandinavian countries, as well as Germany and Austria, legal standards demand that tests prove no blood is flowing into the brain. Tibetan Buddhists, however, believe that after the dying person's breathing stops, "inner respiration" continues for "the length of time it takes to eat a meal"—about another twenty minutes. It's during this period that the process of conception reverses itself. The father's "white and blissful essence" migrates from the dying person's head toward the heart, dissolving anger as it goes, giving the person a clarity of awareness and the experience of "whiteness." At the same time, the mother's "red and hot" essence migrates upward from the abdomen, where it has resided; this dissolves desire, giving the dying person the experience of bliss and of "redness." Meeting at the heart, the male and female essences enclose consciousness between them. The dying person experiences blackness. The mind is freed from all thought, as well as from delusion and ignorance. The dying person then experiences the great state of insight known as Ground Luminosity or Clear Light.

To the Tibetan Buddhists, death does not abnegate conscious-

ness. By dissolving the roadblocks erected by anger, desire, and ignorance, death instead carries one to ever more subtle levels of consciousness until one reaches Buddha nature—what other religions
call the self, or the essence of mind, or God—the subtlest level of all.

Mara slept fitfully, half slouched, half lying on her spot on the
snow bench, Adrian no longer beside her. She dreamt, as many
climbers do when afflicted with the periodic breathing of extreme
altitude, of swimming underwater. As dawn arrived and the first dim
light appeared through the snow window, she saw that Linda was
curled up on the floor of Adrian's chamber, as if to keep her friend
company. There was the sound of rustling in the main chamber as
someone stirred; a dark form that looked like Becca crawled down
the entrance tunnel to the door.

She carefully pushed back the snow block a crack. Suddenly she
pushed it back more.

"It's clear!" she shouted. "The wind's stopped! The sun's rising!"

The women, one by one, crawled from the entrance tunnel and
stood up in the still dawn. They staggered slightly, unused to being
upright, weakened by lack of water and food. They could almost feel
the warmth of the sun about to touch them. The sky glowed pink
and orange over the glacier-draped peaks of the high Himalaya to
the east, while below them Annapurna's slopes, tinged pink, rolled
and humped and fell off toward the 2-mile-deep valley where Base
Camp waited in the lingering pool of night's shadow. To their left,
just past the entrance of the snow cave, the sculpted wavelike form
of the cornice rose slightly and then dropped the thousand feet to
the crevassed surface of the glacier. It was as if their vision now had
suddenly cleared of the frigid, wind-driven mists of two days and
two nights of snowy dreaming.

"If we leave now, we can be back at Camp Three by early afternoon," Becca said, "and Base Camp by tomorrow."

No one spoke. Everyone knew what everyone else was thinking.

"What about Adrian?" Linda finally said.

"I don't know what else we can do," Becca said quietly. "There's no way we can get her down the ice faces, and even if we could, it would take days of work, and the monsoon-season storms will surely be here by then." She swept her hand past the ice-capped peaks and pink sky before them. "Look at this incredible place. I think she'd be happy here. I know I would be."

And so it was agreed to leave her in the snow cave. They crawled back in—already to Mara it seemed impossibly small and claustrophobic—and packed up their gear, anxious to get moving before another storm arrived and yet anxious not to seem as if they were abandoning Adrian. Becca then summoned the team to crowd into Adrian's chamber. Mara could see Adrian's face in the dim light. It appeared grayish white and dull, as if there were no vital forces inside her to reflect the natural light.

Looking at her lying on the snow bench, Mara thought how this peaceful resting spot was preferable to what would have happened to Adrian had she died at home. An autopsy of her brain would have showed edema and hemorrhaging of the tissues, followed by a trip to the funeral home, and perhaps the care of a mortician who would inject formaldehyde or wood alcohol into her veins to preserve the flesh. She'd probably end up in a steel box buried in the earth, with the mistaken notion that it would isolate her from tiny living creatures that might consume her flesh. In fact, there are plenty of bacteria already in the box to do the job.

Instead of trying to shield her body from the earth's natural process of decay by closing up her remains in a box, the Tibetan Buddhists would have cast her body as widely as they could through the world. After death, they'd leave her carefully undisturbed for three days, until her consciousness had the chance to fully leave her body, and then her body would be carried to a special spot in the mountains where monks would chop it into pieces, mix it with barley flour, and let the great soaring birds of the Himalaya descend to feast on it—what the Tibetans call a "sky burial."

Only the most highly trained and prepared individuals are not overwhelmed by the Ground Luminosity when it arrives at the moment of death and are able to use it as the great opportunity for liberation. The majority of dying people return to this endless cycle of death and rebirth and suffering called samsara. The Tibetans believe that Adrian's consciousness, having left her body and not attaining liberation, would linger for forty-nine days in the "bardo of becoming"—the Between, a kind of airport transit lounge for consciousnesses en route to their next lives, as one well-traveled Tibetan master has described it. Her past habits and acts have had consequences—known as her karma—that would buffet her this way and that, finally pushing her toward her next mother and father and the womb of her next life: anything from a cockroach to a bird to a human, who can continue to work toward liberation, to a demigod.

Mara studied her lying cold and lifeless in the dim light of the snow cave and wondered where Adrian might be now. Or maybe she was nowhere and would never be anywhere again. Who knew? Who would ever know? But it was nice to think that even now Adrian was somewhere out there finding her way to another womb. It made Mara feel that Adrian wasn't lost to her. That she would meet her again.

Becca said a short prayer, a kind of nondenominational hope that Adrian's spirit and love for life and for the mountains would inspire others. She had died in the mountains—a place that she loved—and here her body would remain.

"Does anyone have anything else to add?" asked Becca when she'd finished.

"Just a moment," Linda said.

Mara heard the sharp crackle of cold paper.

"I'd like to leave these fig bars and tea bags with Adrian to give her sustenance wherever her next journey takes her," Linda said. She leaned forward and placed the food beside the bench.

Mara felt herself choking up. She remembered the half of a fruit-and-nut chocolate bar, wrapped in golden foil, that she'd tucked away in her parka pocket, saving it for the descent—the last that remained of her food. She pulled it from her climbing suit's inside pocket and gently placed it on the little pile of food Linda had made. Becca then offered up to Adrian her lip balm and sunscreen. Finally, when it came to Gayle's turn, she reached inside her turtle-neck shirt and lifted over her head a gold chain with a stone on it.

"I have no food left to offer, but I'd like to give Adrian this—a ruby that my grandmother gave me when I was a little girl. It's been my good-luck charm all these years, and I'd like Adrian to have it."

She gently placed the loop over Adrian's head, resting the stone on the breast of her climbing suit, and turned to crawl from the chamber. Mara lingered, wanting to be the last out. She gave Adrian a long hug—her friend's body stiff and cold and small beneath her—and then rose to her knees and with her ice axe smashed out the snow window she had built. Bright daylight flooded into the chamber. Kneeling closer, she looked out. Through the opening—the view that Adrian had from her snow bench—Mara could see the deep blue sky and the cliff falling away. The other white peaks of the high Himalayan crest lay below her, and beyond them she saw at last the dry barren hills of the enormous Tibetan Plateau: colors in shades of gray, green, blue, brown, of dry earth and rock, peaks capped here and there with snow. It looked as mysterious and alien as the surface of a separate planet. From this height she could see the luminous layer of air that hugged the hills—the thin veil of atmosphere that wrapped the earth. Here and there a spirelike peak projected up through it, poking through the veil toward the sunlit, blue-black depths of space. She was amazed how thin this layer was that sustained life. She could see from this perch that the Tibetans lived just below the margin between earth and space. No wonder they cherished the existence of even the smallest living thing that managed to survive.

She pulled herself away from the view and considered the dimensions of the window. If any bird wanted to enter the chamber, griffon or kite or vulture or sparrow, there was ample room to fit through the opening. Satisfied with her work, she crawled out the entrance tunnel into the bright sunlight. With the others' panting help, she pushed the snow block in front of the entrance and packed snow into the cracks to seal it. Then the women tied into their ropes, shouldered their packs, and, breathing hard, slowly started down the mountain.

CHAPTER FOUR

THE COLD HUG

OF THE

WHITE SPHINX:

AVALANCHE

All morning, under a cobalt blue sky, the three of them had been snowshoeing with their snowboards slung across their backs up a high ridge in Utah's Wasatch Mountains, a range that offers some of the world's deepest, lightest, and therefore best powder snow. Moist air from the Pacific Northwest churns over the warm Great Salt Lake, then climbs a vertical mile up the Wasatch's wall of 11,000-foot stony peaks into the cold, thin air, releasing its moisture on the slopes below in a thick blanket of feathery flakes. These peculiarities of geography and weather make the Wasatch a mecca for snowboarders and skiers, who travel from throughout the world to the range's winter resorts. Others seek the vast, wild, unregulated expanses of federally owned lands that lie beyond the boundaries of the patrolled ski areas. This is the beyond-the-bounds mountain terrain that its aficionados—like the three snowboarders—know as the backcountry.

Leaving the trailhead at the highway just after dawn, they had slogged up through the aspen groves of the range's gentler lower slopes and entered the shady fir forests of the middle slopes. It was a little past midmorning when they'd run into a Forest Service ranger snowshoeing his way down the trail.

"You kids headed up to the bowl?" he'd asked.

"That's where we're headed," said Jeremy, who was in the lead.

"Be careful up there," the ranger said, eyeing the seven silver earrings that ran like rivets up Jeremy's left ear. "I was just taking a snow survey, and it doesn't look very stable."

"We'll check it out," said Jeremy.

He started snowshoeing past the ranger. But the ranger remained standing in the narrow snow-packed trough of the trail and didn't move to let him pass.

"You better know what you're doing," he said.

Jeremy kept walking forward, kicking three quick steps into the snowy wall of the trail's trough, past the ranger.

"You're blocking the trail, man," Jeremy said, and kept kicking his way up it. Liz and Dougie followed. The ranger glared at them but said nothing. Then he adjusted the straps on his pack and slowly started down again.

They did in fact know what they were doing. They'd attended seminars on avalanche avoidance and wilderness survival, logged countless trips into the backcountry, spent days at a time camping and boarding in high-mountain bowls similar to the one that hovered above the treeline, over their heads, like a great white rising moon. Still, it seemed that they were always tangling with the authorities—the authorities at ski areas who denied them use of the lifts or told them to slow down or get lost, the authorities in the Forest Service who wanted to regulate the use of the backcountry, the parental figures at home. One of the great joys of boarding the backcountry, usually, was that out there lay a whole world free from authority. It didn't discriminate against people simply because they had spiked hair or a pierced tongue or an attitude. The backcountry distributed its joys and its hardships equally, without discrimination.

By late morning, sweating in the bright high-altitude sun, the trio finally arrived at the ridge top. The folded white peaks of the

Wasatch Range stood all around them, some slightly higher, some slightly lower. The great bowl opened at their feet, treeless and white like a giant amphitheater, its rim capped by a 10-foot-high snowdrift called a cornice. Standing back from the cornice's dropoff, they removed their snowshoes from their feet, plunging thigh-deep into the soft snow. They pulled their portable shovels from their packs and dug a pit deep into the snow to study its walls for weak layers in the snowpack. Near the bottom of the pit, they spotted a layer of crumbly snow with a sugary texture.

Jeremy squatted in the pit and picked up a fistful of the snow in his gloved hand. He squeezed his fingers closed. Then he opened it again. The sugary particles of snow refused to stick to each other, pouring from his palm like a fistful of sand.

"It looks kind of loose," said Jeremy, "but on the way up I didn't see any evidence of recent slides in the other bowls around here, so I think we're okay."

He was referring to the telltale scars on steep, snowy slopes and the heaps of churned-up snow at their bases that marked the path of a recent avalanche. Jeremy had more experience in the backcountry than Liz and Dougie, and they deferred to his judgment. The threesome replaced their shovels, strapped their snowshoes to the backs of their packs, and buckled their soft, calf-high boarding boots into the stiff plastic cuffs of their snowboard bindings. The boards themselves resembled short, very fat skis bent up on both ends—a kind of hybrid between a ski and surfboard. They double-checked their avalanche beacons—small radio transmitter-receivers strapped to chest harnesses beneath their parkas. These devices emitted a signal on an internationally set frequency, 457 kHz, that, like the emergency locator beacon of a downed airplane, could be picked up by the radios of others in the group or of other rescuers if an avalanche buried someone. The trio hopped awkwardly toward the lip like frogs nailed to surfboards. The bowl spread out below them, dropping over a thousand feet in elevation—a great symmetrical scoop that an ancient glacier had chewed and ripped from the solid rock of the ridge's bulk that was now blanketed with a thick, white, luscious coat of powder snow.

It was Dougie's turn to make first tracks. The bowl was so big and so steep and filled with so much empty space, it almost made him dizzy to look down into it.

"Well, do you plan to stare at it all day or are you going to board it?" asked Jeremy, reaching down to tighten a buckle on his binding.

"Give me a minute, will you?" Dougie replied.

"When it's his turn for first tracks, he can think about it all day if he wants," said Liz, standing up for Dougie. "That's the way we've always done it."

Jeremy backed off. "I know, I know. It's just that this bowl looks so *sweet*, I can hardly *stand* it!"

"I wish Cat were with us," said Liz, looking at the surrounding peaks, the pinnacles of sharp black rock protruding from the wind-sculpted waves of snow. "She'd love this place."

"I don't know," said Dougie distractedly, still peering down into the long, sweeping open slopes of the bowl. He could feel his heart pumping in his chest, his breathing coming faster. "She really hasn't been that into boarding since Kayla was born."

For three years, Dougie and Cat had been living together in a small rental cabin down in the canyon, waiting tables or working as lift attendants at the ski area as the season and their wallets demanded. With Liz and Jeremy and two or three other boarders, they'd hung out as a group, defying their parents' entreaties to go to college or get real jobs or do something besides snowboard all winter and mountain-bike all summer. Then Cat got pregnant. Kayla was now ten months old—just walking. Dougie loved her—loved to play with her, make faces and watch her laugh, teach her to walk. But some days he had to get out of the little house with its smell of dirty diapers, its crusted puddles of baby food on the kitchen counter, the bouts of crying, the attention that Cat gave to the baby with little left over for him or for the things they used to like to do together. Without boarding, he'd go crazy with claustrophobia. Until he'd discovered boarding, he *had* been crazy. He'd hated his parents—both sets of them. He had been tossed out of school, dispatched to a series of psychologists and counselors and, finally, probation officers. It was boarding that finally gave him a place in the world that felt like his own.

"Whaddaya think, Jer, straight down the middle, or lean a little to the right?" Dougie asked, still studying the bowl.

"I'd say stick to the right," Jeremy answered. "There's a kind of hump in the middle that could really drop off on the other side."

"You think it'll hold?" Dougie asked, more to himself than to the others.

"We'd sure as shit better hope so," said Jeremy. "Look, if you don't want first tracks, I don't mind going ahead."

Dougie looked down into the bowl again. The smooth slope revealed nothing, yet somehow contained all the complexity and random chance of the cosmos. The trillions of snow crystals, each with their own design, were stacked on the slope in an infinite number of possible combinations, adhering to each other—or not—depending on temperature and humidity at their countless points of contact. The slightest extra weight on their surface—the weight of a snowboarder, for instance—could send them tumbling down. Or the slope could hold. Would it avalanche or would it not? You could pose the question, but the slope—smooth, blank, white—would keep its mysterious silence. It was like the Sphinx, the precise alignment of its trillions of snow crystals holding the answer to the riddle. It was like gazing up at the infinite stars of the night sky. Dougie could ask what it intended—what it *meant* for him—but though he could make an educated guess, he wouldn't ultimately know for sure until he tried it.

Dougie pulled his red-lensed goggles from his purple pile hat down over his eyes and cinched up the snow cuffs of his gloves. Certainty wasn't what he valued most in life. If it was, he would have stayed down in the valley and punched a time clock five days a week at a normal job.

"I'll do it," he said.

"Yeah, Dougie!" called out Liz.

"Nail it, dude!" shouted Jeremy.

For a long moment he stared over the edge of the cornice, visualizing his landing and the turns he'd make after it. He breathed slowly and deeply, calming himself. If there happened to have been an EEG machine sitting on top of the cornice and its electrodes were adhered to his skull, the computer screen would show a sudden surge from his left hemisphere in alpha brain waves—the waves, of about 8 to 13 cycles per second (8 to 13 Hz), that correspond to a calm, meditative state. This same surge in alpha waves has been recorded by researchers in archers who are about to deliver an arrow accurately to the bull's-eye, and some athletes are now learning to train their brain waves to put themselves in the "zone" of peak athletic performance. Dougie made three big exhalations and jumped his board so that it pointed straight at the edge. He gave two little hops, then a big one that launched him, plummeting, over the brink.

"Yaaahoooo!"

Dougie heard the quick whistle of wind in his ears as he dropped off the cornice. In the air, he concentrated on keeping a balanced poise—arms out, knees slightly bent, eyes focused on his rapidly approaching landing zone on the snowy slope below the cornice . . .

Whump!

He landed perfectly—"stomped" the landing, as the boarders said. He absorbed the tremendous force of the impact—165 pounds of body weight dropping from the equivalent of a second-story window—by doubling over at the torso and deeply bending his knees. The board punched a divot into the soft layer of powder snow, and an instant later Dougie sprang from his crouch and popped the board up to the surface. Suddenly he was shooting downhill.

He banked into a swooping right turn—his toe side, meaning the toes of his bindings were aligned toward the right side of the board. The board was skinnier in the middle than at the ends; as

Dougie leaned it up on edge to make his turn, it flexed like a bow and carved a graceful arc. Clots of the soft snow flew up and splattered against his red parka and black pants and his face like a cold, wet, delicious brush.

"Yeah, carve it!" he could hear Liz yell from above.

"Make those pow-turns!" Jeremy shouted.

He rocketed out of the first turn, almost airborne, and leaned left—his heel side—and the board sliced down through the second turn, surfacing again as he finished the arc, the steep slope of the bowl falling away beneath him, and he was soaring and plummeting all at once. The feeling was one of utter freedom, the wind sharp on his face, the powdery snow whipping up, the sky a deep winter blue, the bowl huge and untracked below him. He loved it. His body and board played with the implacable pull of gravity, back and forth like a violin bow playing the giant resonating board of the mountain. His brain waves agilely shifted back and forth between frequencies as circumstances demanded, jumping from the quiet alpha up to the normal waking frequency called midrange beta, and above it to states of higher alertness as tricky bits of terrain demanded, and back again to quieter alpha, with the quick and easy shifting up and down that has been measured in the brain waves of B-2 test pilots as they move from easy to difficult tasks and back again while flying their jets.

Three turns. The wind whistling in his ears.

Four turns. He'd found the rhythm.

Five turns. Into long, ripping arcs with a screaming plume of powder trailing behind him like some alpine comet.

Six turns. There was the hump . . . lean right to avoid it.

Seven turns. Not far enough . . . he was headed for it . . . no panic . . . he could see below . . . an easy drop, though the slope steepened below . . . a little air time, was all.

And suddenly Dougie was airborne. He grabbed the heel side of the board with his left hand in the stylish maneuver known as "method air." There was a moment of suspension, a fleeting sense of soaring—birdlike, poised, confident—out over the great expanse of the bowl.

He landed, stomping it again, another perfect landing.

"*Yooooowww!*" he shouted, hoping Liz and Jeremy could hear him on the cornice above and know what a great run he was having.

Eight turns. Something odd out of the right corner of his eye . . . a long, blue-tinged crack in the snow.

Then a sound and the snow reverberated with a hollow noise like a giant hand knocking on a giant pumpkin.

Nine turns. The snow all around him was moving, breaking into thick slabs.

Another shout, his own, one only he could hear.

Already he was instinctively dropping into a low crouch, arms extended for balance—the instantaneous response to a sudden sound or light or surprise known as the "startle reflex." His eyeballs spun, reflexively searching for the source of the danger. The adrenaline pumped into his system. His brain waves screamed up to the high-alert, high-anxiety zone—known as high beta—over 30 cycles per second. The danger was all around him: a slab avalanche, big, soft plates of the new snow fracturing into chunks and falling away down the slope like an old building dynamited for demolition that drops to earth in a sudden cloud of dust.

Dougie swung his board to the right, toward the far edge of the crumbling mass, riding the churning snow like a surfer in the foam . . . for one second . . . two seconds . . . three seconds . . .

A 3-foot-thick slab of snow slammed him from behind, knocking him off his feet.

Onto his back . . . he could feel the avalanche gaining speed, churning snow all around him . . . 40, 50, 60 miles per hour . . . tumbled over . . . submerging . . . swimming with his arms to stay on top . . . feet pinned by his board . . . tumbled over again . . . a breath of air mixed with powder snow . . . like swimming in the foam of a breaking wave . . . wondering when it will stop . . . wondering if it will slam into trees or rocks . . . it was slowing . . . swim hard for the surface . . . one hand over the mouth to create an air pocket . . . shove the other to the top to signal the others . . . the snow suddenly turning heavy and dense and dark . . .

    . . . and then it stopped.

There was a famous holy man in Pondicherry, India, by the name of Yogiraja Vaidyaraja, also known as the Burying Yogi. As proof of his devotion, the yogi routinely climbed into a small box and let himself be buried in the earth, remaining there for several days. In 1973, a team led by U.S. biofeedback researcher Elmer Green traveled to Pondicherry, constructed a box 3.5 feet by 3.5 feet by 5 feet, equipped it with a glass observation window, and sealed the box with wax and polyurethane foam. The box was so airtight that a candle placed inside died out for lack of air after an hour and a half. After the researchers had hooked him up to a battery of electrodes to measure his vital signs and brain waves, the yogi climbed in the box and it was sealed. As Vaidyaraja assumed the lotus position, Green's observations began. The equipment measured that the yogi's brain waves immediately switched from high-frequency beta waves, corresponding to a state of alertness and wakefulness, to the long, slow alpha waves corresponding to calmness and a meditative state. His heart rate dropped by half, and his breathing slowed to a mere one-third the normal waking rate—4 breaths per minute. The yogi remained in this suspended state for a full eight hours before he asked to be removed from the airtight box, complaining of electric shocks received from the equipment.

    The yogi would have made a good avalanche victim—or rather, a good avalanche survivor. Snow is mostly air—anywhere from over 95 percent in fresh, fluffy, undisturbed snow to 60 percent or less in the compacted snow of avalanche debris. The buried victim can suck in the air trapped between tiny snow particles—at least for a short while. This assumes, of course, that he or she has not been killed by the violent tumbling of the avalanche or slammed into trees or rocks or tossed over a cliff, that the churning snow has not clogged his or her air passages, and that the weight now piled on top of the victim has not constricted the chest too much to take in a breath. Even if the victim has survived all this and is breathing beneath the snow, asphyxiation is likely to come soon. Carbon dioxide from the lungs' exhalations will saturate the area around the victim's face, and he or she will begin to suffer from hypercapnia—excess carbon dioxide in the blood—as well as the lack of oxygen in the blood known as hypoxemia. Meanwhile, a thin mask of ice—a kind of alpine death mask—freezes near the mouth and nose due to the victim's warm, moist exhalations, and blocks the airflow from the snow to the victim's lungs.

The statistics of avalanche survival are daunting and precipitous: After fifteen minutes' burial under the snow, the victim has a 92 percent probability of survival. Only twenty minutes later—after thirty-five minutes' burial—that probability has plunged to 30 percent, and after a little more than two hours, the chances for survival are very slim, only 3 percent.[1] Some very fortunate victims, however, have stretched the odds to incredible lengths. In 1955, a twenty-five-year-old Swedish hunter named Evert Stenmark survived burial in a small avalanche for eight days, having melted and carved out an air pocket, and subsisting on a diet of raw ptarmigan, ski wax, meltwater, and dwindling hope while search parties looked for him in vain. A big fan of movies, Stenmark had saved in his wallet every ticket stub of every movie he had ever attended. His brother finally spotted the red stub of a ticket from Stockholm's Black Cat theater

---

1. Figures from "Avalanche Survival Chances," letter by M. Falk, et al., in *Nature*, 1994, 368:21. Also from "Respiration During Snow Burial Using an Artificial Air Pocket," by Colin K. Grissom, M.D., et al., *Journal of the American Medical Association*, May 3, 2000.

that Stenmark had extracted from his wallet, wired to a tree branch he'd discovered beneath the snow, and poked above the surface.

Such are the strange vagaries of avalanche survival. Most people's chances of being caught in an avalanche are extremely remote—unless one happens to be in the wrong place at the wrong time. By one estimate, a hundred thousand avalanches tumble down U.S. mountains each year—the vast majority of them in the remote ranges of the West such as the Rockies, Sierras, and Cascades—and tens of thousands more fall each winter in Canada and Europe. Together they kill only about forty people annually in North America and about one hundred in Europe. While U.S. avalanches mostly claim recreationists as victims—the numbers climbing in recent years as more skiers, snowboarders, snowshoers, and snowmobilers explore the wintry backcountry—it's the villagers in high valleys of the Alps who have lived for centuries under the constant threat of avalanches. They've developed countless strategies to defend themselves: They maintain forests and build fencelike barriers to anchor the snow on steep slopes, place their buildings with a close eye to avalanche run-out chutes, cover roads with shedlike roofs, and even construct their churches and houses with heavy, prowlike walls to deflect the onrush of snow like a ship plowing through the sea.

Unlike the sea, an avalanche's power is instantaneous, the more stunning for its suddenness, reminding us that the ground we live on and the mountains themselves—those very symbols of stability—are not static but in their own way alive and ever-changing, as is the snow that covers them. When it does release, the thundering rush of snow can flatten houses, buildings, and forests; it can rush over the top of smaller mountains like a great white wave or bury an entire valley floor. Avalanches have been clocked in Japan's mountains at speeds of 230 miles per hour, have been measured in Alaska as traveling a distance of eight miles, and have killed—in the case of a Peruvian ice, snow, and mud avalanche that in 1970 broke off an Andean mountaintop and dropped thousands of feet, tumbling miles through a populated valley—twenty thousand people.

Some of the most famous victims of avalanches include Hannibal's soldiers and battle elephants, decimated in 218 B.C. by avalanches as the battalions crossed the Alps from the north on their way overland from Spain to attack Rome. "Snow falling from the high summits engulfs the living squadrons," wrote the epic poet Silius Italicus. Not far from that spot, two German hikers in 1991 discovered the five-thousand-year-old body of a hunter thawing from the ice of a glacier, soon to be world famous as the Ice Man, or Otzi (after the Otztal Alps), and possibly the victim of a prehistoric avalanche. Prince Charles narrowly escaped burial in an avalanche during a 1988 ski outing in the Alps when he and his party, including a Swiss mountain guide, ventured into high terrain that had been deemed hazardous, and one of his closest friends, Major Hugh Lindsay, a former equerry for the queen, perished in the slide. Hundreds of climbers have died in avalanches in ranges all over the world, including Alex Lowe, considered by many to be America's best alpinist, who was buried by an enormous avalanche in 1999 while attempting to ski a mountain in Tibet.

It is not difficult to understand how some of the world's most knowledgeable mountaineers have been caught or killed by avalanches. For all the research that has been conducted into their causes and the physics of snow, avalanches remain one of the most frightening phenomena of the mountains. The lure and beauty of the high mountains in winter is almost irresistible for some, and yet it is nearly impossible to go there without encountering some degree of avalanche risk. Each year, hundreds of thousands of skiers, climbers, and snowboarders willingly take that risk and set off into the wintry backcountry. And if they do decide to ski or board a steep slope of untracked deep powder—that blank and tempting mystery—it is truly a leap of faith. It can either reward them with an ecstatic and heart-thumping run or finish them with a slow suffocation beneath thousands of pounds of snow.

There was a faint dawnlike light. There was no sound. The weight of the snow pressed on him. He tried to take a breath; his chest expanded slightly and was squeezed by the settled snow around him that already felt as hard as a concrete shell.

He smelled wet leather. The palm of his glove was pressed close to his face. It formed a small air pocket. Dougie tried a smaller breath . . . and then another . . . and another . . .

He sensed he was lying on his side, his face turned partially up. He squirmed, trying to move his arms to dig a bigger air pocket in front of his face. They were pinned, utterly, as if in plaster casts bolted to a steel table. He tried to move his legs. Nothing. Only his toes moved inside his boots. He tried to twist his torso. Nothing. It was like being locked inside a cold, dark trunk with his hands and feet bound.

He wanted to scream.

Struggling, he frantically twisted his head from side to side as if to drill and butt his way out. His head moved slightly, the snow crunching in his ears. He'd lost his hat somewhere. He knew it was cold, but he wasn't thinking about cold; hypothermia wouldn't set in for a good deal longer. He wiggled his hand. His fingers moved inside his glove. With his fingers he clawed at the snow in front of his face. The pocket grew an inch or two in diameter. But then a chunk of snow that he dislodged with his fingers dropped into his mouth. He was choking on the snow, a cold cottony mass in his mouth. He tried to spit it out, but it was too big. With his gloved index finger, he managed to snag the chunk and extract it.

He could breathe again, but already his breath was coming harder, the air moist and dense. He felt panic rise up in him, his childhood fear of small, dark rooms, of waking in his bed and choking on the blackness, the night like a pillow held over his face, smothering him, stopping him from screaming out for his mother and father.

"Help!" he shouted, banging his head back and forth. "Help! Get me out of here!"

The snow muffled his shout. The effort made him pant. Someone pushed the pillow harder against his face.

I'm going to die, he thought. *If they don't find me, I'm going to die.*

The panic surged up in him again. He smashed his head harder against the walls of snow, clawing with his fingers at the same time. Now his breath came in short pants, his chest squeezing against the hard shell of snow, his heart pounding at over 100 beats per minute.

He'd been buried for two minutes. The oxygen in his bloodstream had dropped only slightly, from 92 percent saturation—the level a twenty-year-old male would have while breathing normally at about 9,000 feet—to a little under 91 percent. But the carbon dioxide in his exhalations had jumped, and, worse, the carbon dioxide in his *inhalations* had soared by a factor of 100, climbing from about .035 percent of the air in each breath—just a trace amount for someone walking around breathing normal air—to 4 percent carbon dioxide in each breath. This level was still too low to affect Dougie's central nervous system—the limit of human consciousness occurs when breathing air with a 15 to 20 percent carbon dioxide concentration—but it was enough to trigger the carbon dioxide sensors in his brain to tell his lungs to work harder, to breathe more deeply and faster to try to clear his bloodstream of the carbon dioxide. Meanwhile, his brain waves were bouncing around in the 30- to 35-cycles-per-second range of high beta—anxiety, fear, hyperactivity.

He screamed again.

Again the snow muffled his plea, as if the huge mass on top of him remained utterly indifferent to his desperation. He thought of Cat and Kayla sitting on the old frayed sofa down in their cabin in the valley, waiting for him to walk in the door. He'd never see them again. He smashed his head back and forth in a frenzy, as if by butting his skull against the hardened snow he could obliterate the actuality of his entrapment.

He stopped. He was panting hard now but still remained breathless. He wanted to yell and twist again, trying to slam his way out. But if he did, he realized, he'd run out of air much more quickly.

He'd now been buried for four minutes. His blood oxygen saturation had dropped to 89 percent, still not low enough to cause major discomfort, as people can survive with blood oxygen saturations as low as 50 percent. The carbon dioxide he exhaled had filled the air pocket immediately in front of his face and saturated the tiny air pockets between the snow crystals. From that same stale pocket he was now inhaling air that contained 6 percent carbon dioxide, and the concentration was climbing quickly, into the range that impairs mental capabilities. He was beginning to show the symptoms of acute hypercapnia—excessive carbon dioxide in the blood—which range from faster and deeper breathing, anxiety, and some impairment of mental function at levels of 4 to 10 percent to severely impaired brain function at levels of 10 to 15 percent, followed by unconsciousness at 15 to 20 percent and convulsions at levels over 20 percent. Monkeys and dogs breathing an extremely high level—30 to 40 per-cent carbon dioxide—have showed an elevated heart rate and other cardiac irregularities but were able to survive breathing the carbon dioxide–air mixture for many hours, although if they were returned too abruptly to normal air, their hearts tended to go into ventricular fibrillation—severely abnormal rhythms—and they abruptly died.

Had Dougie been wearing a new type of vest that recently appeared on the outdoor-equipment market, he might have avoided some of these problems. Instead, he would have inhaled through a tube in the vest's collar that connected to a mesh-fabric area embedded in the vest. This mesh acts as an artificial air pocket for avalanche victims. He could suck the pure air straight from the snow that lay against his vest front while the carbon dioxide from his exhalations would be vented by a separate tube into the snow at his back. The amount of carbon dioxide he inhaled would not have climbed nearly so high or so quickly as it did breathing from the pocket in front of his face.

But he wasn't wearing a vest like this. The high concentration of carbon dioxide in his bloodstream kicked up his respiratory rate and caused him to pant, but the panting did no good. With every quick, deep breath he simply took in more carbon dioxide. There was only so much room for gases in the tiny sacs of his lungs called alveoli; as the carbon dioxide level rose in them, the amount of oxygen they contained dropped. Dougie was beginning to suffer from what is known as displacement asphyxia.

Liz and Jeremy hooted as Dougie made his great carving arcs down the bowl followed by a contrail of powder. They saw him launch over the hump, grabbing his board in a graceful method air. They heard his "yooowww" of delight as he stomped the landing. Then they saw him suddenly drop into a crouch. His arms shot out for balance. They could see the slope around him fracturing into slabs and fall away, slowly at first, and then with amazing speed. Standing on the cornice several hundred yards above the avalanche's starting zone, they were not themselves in any danger from the slide.

"Get out of it, Dougie, man," Jeremy shouted, "get out of it!"

"To the side, Dougie, to the side!" Liz cried.

But Dougie was down, a black and red speck, and then he was gone, swallowed up in the churning, descending cloud of white.

There was no noise at first; it took a second or two for the sound to travel up to where they stood on the cornice. Then they heard the thunking cracks of the snow, a whoosh, and an enormous roar that reverberated around the bowl.

"Don't lose the spot where you last saw him!" Jeremy shouted to Liz. "I'll watch for him down in the run-out."

Within a few seconds, the plume of snow had swept the 400 or so yards to the bottom of the bowl, spilled out onto the flats in a broad rubbly fan at least the size of a football field, and abruptly come to a stop. The roar continued to echo for a moment after the snow stopped moving. Then it was silent. Jeremy squinted at it. No red. No black. Nothing.

"I have it marked," said Liz.

She pointed to the spot where they'd last seen him—about 50 yards downhill from the hump's right side. They each marked it in their minds.

"We have to be really careful," said Jeremy, "or we'll be buried, too. I'll go first, and when I'm below the hump, you follow."

Jeremy hopped his board over the edge of the cornice, dropped into the bowl, and made long fast swoops down to the hump, sheltering himself below its bulk in case Liz knocked down another avalanche from above. In a minute she had joined him, just above the avalanche's fracture line—a 3-foot-high escarpment in the snow where the slabs had split away. Underneath lay a smooth and even surface that was exposed when the thick outer scalp of snow peeled away.

Within the snowpack, stacked in layers like the leaves of a book, lay an entire history of the winter as it occurred in the bowl. Two months earlier, winter's first big snowfalls had been followed by a cold snap. The warm earth below the snowpack and the frigid air above it caused the snow particles near the bottom to vaporize. The water vapor migrated upward and recrystallized on the underside of the uppermost, colder particles that lay near the frigid air, like frost coating a windowpane. Eventually the entire layer had recrystallized into tiny ribbed pyramids and cups known as "depth hoar" or "sugar snow." No bonds held these loose particles. Then more snowstorms fell on top of this depth hoar, but these newer layers didn't undergo the big temperature difference between warm earth and frigid air, and the particles bonded with each other. First the little arms of the individual snowflakes broke off or receded until there was a roundish particle of ice, and then tiny necks of ice called sinters formed from one particle to the next, like the little knobs holding together a string of fake pearls. This loosely bonded mass of trillions of snow particles and thousands of tons of snow, however, rested on the weak, shaky foundation of the unbonded depth hoar. Dougie's landing—*whump!*—was enough of a shock to send a fracture through the upper layers and set them sliding over the weak underlayer, rapidly accelerating in the great, crumbling plates of a slab avalanche.

Liz and Jeremy side-slipped on their boards quickly down the scalped surface of the depth hoar, straight downhill from the spot where they'd last seen Dougie. The avalanche rubble at the slope's base was like the snow piled up at road's edge by a plow—big chunks interspersed with looser snow, the whole of it already bonded into a solid mass. As they reached the rubble, they flipped open the buckles of their bindings and jumped out of their boards.

There was no sign of Dougie.

"Let's switch to receive," Jeremy said. He, too, was hyperventilating with anxiety. His heart was hammering at 135 beats per minute. Liz and Jeremy yanked off their gloves, reached into their parkas, and pulled their transceivers out of their harnesses. They pushed the buttons that allowed them to receive the signals emitted by Dougie's. As they did they heard a steady *beep . . . beep . . . beep.* His signal was here somewhere. The question was exactly where. Their transceivers had directional antennas built into them, and a glowing red arrow would light up when the transceiver was pointed in the direction of Dougie's signal. Standing in the avalanche debris, they rotated their transceivers back and forth to find the direction of his signal while listening to the steady beep.

It had now been seven minutes. Dougie was climbing a high peak with a heavy pack through deep snow—panting, panting, panting—but didn't have enough air to take in. He dreamt that he'd been caught in an avalanche while climbing the mountain. But he must still be climbing the mountain because he was panting so hard. The carbon dioxide in each inhalation had risen over 7 percent—closer to the range of severe mental impairment. His blood oxygen saturation had dropped to 83 percent. His breathing rate, due to the carbon dioxide he was inspiring, had jumped to nearly 30 breaths a minute—heavy, difficult panting—but he was taking in less and less oxygen. What finally kills avalanche victims, it appears, is severe hypoxemia—lack of oxygen in the blood that sends the body into a kind of downward, self-perpetuating spiral. The lack of oxygen in the blood eventually causes brain function and heart rate to slow. With the slowing heart rate, blood pressure drops. The victim blacks out under the snow. The dropping blood pressure causes even less oxygen to reach the tissues of the heart. The oxygen-deprived heart slows even further. Blood pressure drops more, and still less oxygen reaches the heart. Finally, between 15 and 35 minutes after burial, the heart usually stops beating altogether.

Dougie thought of Cat and Kayla on the sofa again, realized where he was, and felt the panic rise in him again. He woke up, wanting to scream. Then he thought that if he screamed, he would die. He would use up even more oxygen. No, he couldn't scream. He had to do the opposite of scream. He had to be very, very quiet. He had to believe the others were out there, somewhere, looking for him. He began by slowly moving the fingers of his gloved hand. He managed to scrape his air pocket an inch or two larger. Instantly, he felt some relief as fresher air leached from the snow into the pocket. Now he had to find the calm spot inside himself. *Just go with it,* he told himself, like a mantra. *Just go with it. Just go with it.*

Seven years earlier, after he'd been suspended from junior high school, after he'd been arrested for shoplifting, after his parents and teachers had sent him to a long series of meetings with counselors and he'd been diagnosed as hyperactive, an innovative juvenile-court judge had sentenced Dougie to neurofeedback therapy. The therapist, a gentle, middle-aged woman, pulled aside tufts of his unkempt hair, dabbed a bit of conductive adhesive on several places of his scalp, and fixed electrodes to them. He reclined in a soft armchair and watched a computer screen across the small office. A broad highway unfurled toward a pastel desert horizon dotted with distant buttes that reminded him of Arizona. The therapist told him to try to prevent the highway from widening as it unfurled, although she didn't tell him how.

"You'll discover for yourself how to do it," she said.

The highway's width was actually a measure of the frequency of Dougie's brain waves—the electrical pulses that constantly move through our brains and shift frequency for reasons that are still not clear but which correspond to different levels of mental activity or agitation. Neurofeedback researchers have found that those suffering from attention deficit disorder are more or less stuck down in the slow, low-level alpha and theta frequencies near daydreaming and sleep, while those who suffer from hyperactivity spend too much time in the high-level, agitated beta frequencies. The theory of neurofeedback is simply to train the mind to shift more easily between frequencies; it has also been shown to be highly effective in treating epileptics, though the practice is still not widely accepted by mainstream medicine.

Dougie discovered that with a certain, indescribable kind of soft focus, he could quite easily keep the unfurling highway narrow. When he did, the computer made a soft binging sound that indicated he was scoring points—a kind of game—and when he scored enough points the buttes on the horizon erupted in a glorious display of color. After finishing each forty-minute session, Dougie felt relaxed and the world seemed remarkably vivid and clear. After six months of the sessions, his concentration had improved vastly, and he was calmer and doing much better in school.

Now, pinned under the thousands of tons of snow in the semi-darkness, panting, wanting to scream, he thought of those sessions. He remembered that soft focus. He remembered trying not to concentrate on any particular object or thought but on a kind of empty space. Now he imagined that space—all the space inside this cast of snow, the spaces between the trillions of snow crystals, all the space of the universe. He couldn't let himself think of Cat and Kayla playing on the sofa, waiting for him to walk in the door. Instead he had to think of a vast empty space.

In a way, what Dougie was doing was not all that different from the basic meditation techniques of Eastern yogis, who instruct the aspirant to attempt to empty the mind of all thoughts while meditating, and find serenity within that emptiness. If there were an EEG machine and electrodes hooked to Dougie's scalp, they would now have shown his brain waves move out of the high-beta frequencies of extreme agitation. They slowly slackened from around 30 cycles per second down through the lower beta ranges and into the alpha frequency at 8 to 13 cycles per second. There were a hundred billion nerve cells in Dougie's brain. Like snowflakes, each one had a different structure as well as a different function. Unlike the billions of snowflakes that surrounded him, each of which was in contact only with those snowflakes immediately beside it, a single brain cell can potentially communicate by tiny electrical impulse with several million other brain cells. All the scientists in the world could not at this point tell exactly what was going on in Dougie's brain, or in anyone's brain; it has been calculated that the number of possible combination of connections—one definition of the number of possible brain states or "thoughts"—in a single human brain exceeds the number of atoms in the universe. So complex that it makes a supercomputer look as simple as an abacus, the human brain may never truly be able to understand itself.

Dougie's brain was now sending out what are known as synchronous alpha waves, which can be triggered by the imagining of space. In the normal waking beta range, the brain emits somewhat different frequencies from different parts of the brain. But in a state of synchronous alpha, the waves originate everywhere synchronously—"like a smooth hum over all the brain," as one description puts it. It is a deeply relaxed state, one where the entire body seems to let go of its stresses. In this state, Dougie's heart rate now slowed from 120 to 70 beats a minute. His respiratory rate, though still fast, became steadier. He'd entered the realm of the Burying Yogi. He had imagined calmness and empty space, and then he was in it. He was surrounded by space. The snow crystals touched one to the next; his neurons touched one to many, sending up the pulsing electric waves of synchronous alpha, a cycle spinning out into the spaces between the snow crystals, between the stars. He imagined his synchronous alpha waves pulsing out into space, toward unknown receivers thousands of light-years distant, like the mysterious electromagnetic signals received by earth's astronomers from far-off stars and the deepest reaches of the universe.

For the first time since he'd triggered the avalanche, he was not struggling and stirring. He could hear the slow, powerful thump of his heart. For the first time, the snow was not scratching in his ears as he banged his head about in his encasement. He now heard down near his chest the transceiver sending out its signal, a tiny beep each time it launched an electromagnetic pulse out through the densely packed snow.

On the surface, Liz and Jeremy were working their way through the chunks of snow in the bright morning sunlight, watching and listening to their transceivers. There was always the danger of another avalanche coming down on top of them. Stumbling often in the rubble, they moved quickly, knowing that these were the most crucial minutes in an avalanche rescue. Rotating their beacons back and forth, they picked up the direction of Dougie's signal, indicated by the red illuminated arrow. They started walking toward it, spread slightly apart from each other. From a grove of subalpine fir trees at the bottom of the bowl, Liz could hear a crow protesting their presence in the sunny morning calm. Then she spotted something purple 30 feet ahead of her.

"Dougie's hat!" she shouted out to Jeremy. "Here's his hat!"

He ran across the rubble to where she stood, holding the hat.

"Check your beacon!" he said, panting.

They looked down at the small readout on her beacon. It read 12 meters. That was the distance from them at which, according to her beacon's interpretation of it, Dougie's signal originated.

They stumbled across the snow again, following the directional arrow on their compact yellow boxes, the handles of their rescue shovels protruding from the backs of their packs.

The crow flew through the still winter air to another tree.

The numbers on the readout on Liz's beacon dropped: 12 . . . 7 . . . 4 . . . 1 . . .

She stopped.

"Here!" she shouted to Jeremy. "He has to be right here!"

He had already peeled off his pack, stripped off the shovel, and started to dig in the hard snow.

A calmness had overtaken Dougie. He was dreamy again. *If this is dying,* he thought vaguely, *it isn't hard.* He was using only a small amount of oxygen. There was only a small amount of oxygen left to use.

In his left ear, Dougie heard a peculiar crunch. It sounded very close and harsh—not something pulsing vaguely between the stars. Something hard hit his left shoulder, then his head. It might have hurt if he'd cared. It was as if he'd been sleeping. Now somebody was shaking him. He opened his eyes. A black-gloved hand was moving back and forth over his face, brushing away the clumps of tiny crystals of snow.

CHAPTER FIVE

STELLER'S LAST LAUGH, OR

BETTER LIVING THROUGH BOTANY:

SCURVY

Gusts whistled through the rigging and rain thumped on the deck planking just overhead, but down below in his sailboat's small salon, Phil was warm and dry. On the teak-wood table in front of him lay an open book, and beside it stood a glass of the red Bordeaux that his local Seattle wine merchant claimed would be highly drinkable on a long sea voyage, although Phil—whose tastes usually ran more to highly caffeinated soda—was the first to admit he wouldn't have known the difference if the stuff were swill. Still, a glass of claret seemed in the spirit of his undertaking—the beverage that British Royal Navy officers were always knocking back before jamming their ships into the icepack for another doomed attempt at the Northwest Passage over the top of North America. Likewise, Phil could have switched on the salon's 12-volt electric lights but preferred the soft glow of the reproduction brass oil lamp that he'd purchased for $199 at the ship's chandlers before he'd left

Seattle three days earlier. His first discovery on this voyage of self-discovery had been that nautical atmosphere did not come cheap.

With its electronic beep, the marine radio signaled the start of another cycle of the weather report. Phil looked up from his book.

"Small-craft warning in the Strait of Juan de Fuca for tomorrow, with southwesterly winds of twenty-five to thirty knots and swells six to eight feet. . . ."

It was the same forecast he'd heard all afternoon. This storm system would be followed immediately by another, lingering at least until Sunday. The question for Phil was whether to chance a crossing of the Straits in the morning or wait two, three, or even four days for the wind and waves to drop. Three or four days! That was an eternity in the world where he'd been operating—long enough to start up a company and then sell it.

He went back to his book. He was finding it hard to put down, despite his anxiety about the weather. Georg Wilhelm Steller's *Journal of a Voyage with Bering, 1741–1742*. Vitus Bering was the aging Danish navigator commissioned by Russia's Peter and Catherine the Great to sail east from Siberia and locate the unknown northwestern coast of the North American continent. He is now memorialized in the names of the Bering Sea and the Bering Strait, the latter separating Asia from America. The journal's author, Steller—who served as the physician, mineralogist, and naturalist on the voyage, as well as Bering's cabinmate—was a young, rigorous German scientist who held nothing but contempt for the ship's officers and crew. The feeling was mutual. Unlike so many gilded explorers' chronicles, Steller, embittered by the treatment he received, pulled no punches—especially when it came to describing the officers' incompetence and misjudgments and the astounding series of disasters that these shortcomings helped trigger.

Phil's girlfriend, Myrna, had given him Steller's journal just before he set sail. Or rather, his ex-girlfriend. Several months earlier, when he began planning the voyage, he'd asked her to accompany him. She'd thought about it for a while; in the end, she'd turned it down—and turned down Phil, more or less, along with it. Things just weren't going to work between them, she'd said. She wanted a settled, predictable life. She wanted a nice house, a vegetable garden, a regular job with regular hours, and children. He was too restless for that—or so she'd said. He had protested, claiming that he wanted that life, too—someday, although in truth he wasn't sure. But first he had to make this sea voyage. He had to do this before he did anything else. Couldn't she understand that?

"Yes, I can understand that," she'd said. "Just don't expect me to be waiting for you when you return."

That had been the last they'd seen of each other. But as Phil made his last-minute supply runs, from the chandler's to the grocery store to the bank, with stops in between, he returned to his boat at its downtown mooring to find Steller's journal placed carefully on the deck, along with a small package wrapped in brown paper that bore Myrna's return address. He dropped the book through the hatchway onto his berth and angrily stowed the package below without opening it. He then hauled down the boat's gangway his bags of groceries and cases of Bordeaux, his new foul-weather gear and the spare fittings for the deck, the other sea journals that he'd collected to read on his journey, charts, spare batteries, and dozens of other items.

His plan—admittedly vague—was to sail off somewhere and read the journals and drink the wine until he figured out what to do next with his life. The Seattle software company where Phil had worked had seen its stock climb in value at an astounding rate over the last year; Phil realized it had to stop somewhere, so he bailed out before it did, quitting work and selling the stock. With a chunk of the money, he'd purchased his sailboat—not a huge yacht, but a nice 35-foot sloop that had a fiberglass hull and handsome wood trim. Below deck, a fore cabin in its bow contained a V-berth double bed; moving aft, one encountered a closet-sized space containing a small shower and toilet—or head, as it was known at sea—and still farther aft, in the widest and deepest part of the boat's hull, sat the salon, offering plenty of headroom and a teak dining table that converted to an extra bunk, plus the sailboat's small galley. The boat was also equipped with automatic sail-furling devices that allowed him to man it single-handedly, plus a good radio as well as a GPS device that located his position on the globe via satellite.

He knew something about sailing, having handled small boats at his family's summer cabin on a lake on the east side of the Cascades, although this kind of big-water sailing by comparison was the difference between operating a pocket calculator and a computer equipped with the latest high-speed processors. The yacht broker had assured him that the boat was fully capable of crossing the Pacific to Hawaii and beyond—assuming Phil was bold and skilled enough to handle her. He'd proudly christened her the *Myrna Loy*, after the actress and after his girlfriend, who at that point he'd still hoped would accompany him. His chances of seeing her again, however, hadn't been helped when the shipyard painters misspelled it the *Myrna Lay*.

North, he'd finally figured. He'd sail north. He'd much more quickly find the wilder shores where he could think things over by sailing north than south. If he really wanted, he could sail the Inside Passage all the way to Alaska—staying in the shelter of islands most of the way except for a few big gaps where he'd be exposed to the open ocean. Only three days out from Seattle, he'd reached the first of those gaps—the 25 miles of the Strait of Juan de Fuca, between the southern edge of the San Juan Islands and Victoria, British Columbia. Daunted by the wind and rain that swept down on him, the big swells hammering against the *Myrna Lay*'s bow, the cold rain trickling down his neck, the seesawlike heaving of the deck, he'd sought shelter late that afternoon and anchored off an island in a small cove rimmed by a forest of fir trees.

He was hoping to find the courage in the Steller journal to

weigh anchor in the morning. Phil admitted that, whatever else the Bering expedition's problems, it must have taken a lot of heart to sail a small ship across a stormy, unknown northern ocean to an unexplored coast. Phil's was in part a journey to the unexplored coasts within his own mind—the farther shores of possibility—which were looking a lot farther than he anticipated now that he'd stalled out a mere three days into his voyage.

He leaned over the journal again in the soft yellow light from the brass lamp. At least on board the *Myrna Lay* he didn't have to deal with any warring personalities, unlike on Bering's *St. Peter*. The conflict was apparent almost from the moment she and her sister ship, the *St. Paul*, sailed from Siberia's Kamchatka Peninsula on June 4, 1741. Steller, son of a music teacher, was a self-motivated student who had worked his way as part-time preacher through courses in theology, philosophy, medicine, and the natural sciences at Halle, one of Germany's most progressive universities. After his university education, he'd moved to Russia to try his prospects there and became captivated by the idea of joining the explorations then under way in the Russian Far East and North Pacific. Only thirty-two years old when Bering hired him on and the expedition got started, Steller immediately began dispensing unsolicited advice based on his learned scientific observations to the *St. Peter*'s roughly educated, and mostly Russian, officers and crew.

Sailing across the North Pacific toward America, Steller urged a more northerly heading because floating seaweed and other flora and fauna indicated to him that land lay nearby in that direction. The officers brushed him off.

"You are, after all, no seaman," they told him.

"They, of course," confided Steller to his journal, with a pen he never hesitated to dip in acid, "have been in God's council chamber!"

He noted that by ignoring his advice, it took the *St. Peter*—which soon became separated from the *St. Paul*—an extra six weeks to find land. When they finally did reach the coast of what is today

Alaska, Steller, observing the local water salinity and currents, pointed out to the officers what he deduced was a good anchorage at a river mouth. They told him to forget it.

"Have you already been there and made sure?" they asked him.

When they did drop anchor in a spot of their own choosing, the officers wanted to stay only long enough in America to replenish their water barrels and immediately turn back toward Siberia, not even allowing Steller off the *St. Peter* and onto the American shore with the water carriers to fulfill his naturalist's duties. They finally did concede to his ten-hour frenzy of botanizing and collecting on the new land, then sent him a message. "I was to get my butt on board pronto, or, without waiting, they would leave me stranded."[1]

These conflicts, however, were only differences of opinion over nautical matters; the real problems commenced on the voyage back to Siberia, when scurvy took hold. A few weeks into their return trip across the North Pacific, the crew began to weaken. Bering was bedridden with the disease by the time the *St. Peter* anchored at an island off today's Alaska Peninsula for more water. Steller asked the officers to assign him a few men to help gather plants that he knew would help cure the disease, and he pointed out a spring that offered fresher water than the brackish pond the ship's crew had chosen to refill the barrels. By then, however, the officers—to their own great detriment, it would turn out—had altogether ceased to listen to Steller's unsolicited advice. Steller, equally disgusted with them, on the spot made a vow.

"Although I had already built a hut because of the evening's rain and had wished to spend the night ashore, I decided to go on the ship to bring up, emphatically and with the greatest modesty once

1. Steller's quotations here and elsewhere in this chapter are from *Journal of a Voyage with Bering, 1741–1742*, by Georg Wilhelm Steller, edited by O. W. Frost and translated by Margritt A. Engel and O. W. Frost (Stanford, California: Stanford University Press, 1988). The actual German of this pithy quote in Steller's manuscript reads: *"ich sollte mich nur geschwinde nach dem Fahrzeuge paken."*

more, my opinion about the unhealthy water and the gathering of plants. But when, concerning the water, I saw myself spurned and rudely contradicted and heard myself ordered to gather the plants as if I were a surgeon's apprentice subject to their command, and the matter that I recommended affecting their own interest, health, and lives was not deemed worthy of the work of one or two persons, I regretted my good opinion and resolved, in the future, to look after the saving of myself alone, without the loss of one word more."

Phil heard the beep of the weather radio, starting its cycle again. He looked up from the journal and listened. Still the same report. Still the heavy winds, rain, and 6-to-8-foot swells. He could feel the gusts through the rigging sway the *Myrna Lay* and shift her bow about on the anchor line. He hoped the anchor would hold in the cove's rocky bottom. He didn't have much experience with anchors. In fact, as he'd motored into the cove, he'd had to pull out his sailing manual and frantically skim over the section on anchors and anchoring before he ran to the bow, heaved his anchor over the side, and paid out what the manual said was the proper amount of line.

He poured himself another glass of the Haut-Médoc. He looked about the salon—the brass lamp, the wood paneling, the shelf of books with the wooden bar across to hold them in place when the boat heeled over, the little galley, the food lockers. All Steller's talk of food made him hungry. He stood up and went to the food lockers. He unlatched the uppermost one and surveyed its contents: two large bags of barbeque potato chips, five cans of ravioli, eight packets of creamy chicken ramen, three 2-liter bottles of diet cola, an 8-ounce bag of gummi bears. He unlatched the bottom locker: a jar of peanut butter, a jar of jelly, two loaves of white bread, tortilla chips, a 2-pound bag of Seattle's finest coffee, a 5-pound bag of sugar . . .

It was then that Phil realized he hadn't stowed aboard the *Myrna Lay* a single piece of fresh fruit or a stalk or leaf of vegetables.

In the late 1400s, with the advent of new ship designs and navigation techniques, and much longer voyages in search of sea routes to the Orient, scurvy—"the explorer's sickness"—suddenly became epidemic among ships' crews. Typically it struck after ten to twelve weeks at sea. "Many of our men fell ill here," the Portuguese explorer Vasco da Gama wrote in 1497 as his expedition rounded the southern tip of Africa en route to the Orient, "their hands and feet swelling, and their gums growing over their teeth so that they could not eat." The men were saved by the timely appearance off Mombasa of a boatload of Moorish traders carrying oranges—"better than those of Portugal"—that quickly firmed up their gums.

This pattern was repeated in countless expeditions that followed. After three or so months at sea, the crew succumbed to scurvy and rapidly started to die unless they made landfall, where a few days' worth of fresh food "miraculously" rehabilitated them. One sharp-penciled Portuguese paymaster calculated in 1634 that in the previous five years, less than half of five thousand soldiers making the outbound voyage between Lisbon and India survived the twin catastrophes of shipwreck and scurvy—known to them as *amalati de la boccha*, or "the curse of the mouth." One leading theory was that bad air caused scurvy, especially the bad air belowdecks, although the Spaniards on transpacific voyages believed they fell victim to a foul and subtle wind blowing off California's Cape Mendocino that triggered—and perhaps still does—"some corruption of the bad humors."

Though individual sea captains and ship's physicians apparently understood that fresh fruits and vegetables prevented or cured scurvy, a remarkable amount of confusion persisted about it for nearly three

centuries. It is almost inexplicable why a sure cure, so readily available in nature and so well known to many native peoples, was not
widely prescribed by the European sea powers hundreds of years
sooner. Because there was no standard prescription for scurvy for so
long, some two million sailors lost their lives between 1500 and 1800,
according to calculations made from nautical records.

There were ample clues to scurvy's causes and cures. In the
1530s, a French expedition under Jacques Cartier searching for a sea
route across North America wintered in the icebound St. Lawrence
River, and by mid-February scurvy had laid low the crew. Finally
only three or four men—including Cartier, who must have nourished himself from a private food cache—possessed the strength to
move from their bunks. Twenty-five of a party of 110 had died.
Cartier ordered an autopsy on one of the victims, twenty-two-year-
old Phillip Rougemont of Amboise, in the hopes of finding a hint
about how to stop the disease from killing the entire crew.

"It was discovered," wrote the expedition's chronicler, "that his
heart was completely white and shrivelled up, with more than a jug-
ful of red date-coloured water about it. His liver was in good condition but his lungs were very black and gangrened. . . . His spleen for
some two finger breadths near the backbone was also slightly affected, as if it had been rubbed on a rough stone. After seeing this
much, we made an incision and cut open one of his thighs, which on
the outside was very black, but within the flesh was found fairly
healthy. Thereupon we buried him as well as we could."

Cartier wouldn't let the local Indians near his ship for fear
they'd see the expedition's vulnerability to attack, but it was those
same Indians who had a cure. Outside the ship one day, Cartier ran
into Dom Agaya, an Indian who, when Cartier had first met him
some days earlier, was suffering from his own case of scurvy. When
Cartier asked how he had been cured, Dom Agaya and two women
from his tribe showed the captain how to cut branches from the
"annedda" tree, grind up its bark and foliage, boil it in water, and

drink the infusion every two days, applying the dregs to scurvy-swollen legs.

"The Captain at once ordered a drink to be prepared for the sick men," wrote the chronicler, "but none of them would taste it. At length one or two thought they would risk a trial. As soon as they had drunk it, they felt better which must clearly be ascribed to miraculous causes; for after drinking it two or three times, they recovered health and strength and were cured of all the diseases they ever had. And some of the sailors who had been suffering for five or six years from the French pox [syphilis] were by this medicine cured completely. There was then such a press for the medicine that in less than eight days a whole tree as large and as tall as any I ever saw was used up."

One modern investigator concluded that the mysterious and potent "annedda" tree probably was the white cedar. An expedition a few years after Cartier's journey carried samples of the white cedar back to France, where it received the name *arborvitae*—"tree of life"—that is so familiar to gardeners these four and a half centuries later. Its powers to cure scurvy, however, apparently were soon forgotten. Not ten years after the Cartier expedition, another French party wintering in nearly the same spot on the St. Lawrence lost fifty out of two hundred to scurvy. A few decades later, wintering on the St. Croix River in today's Maine, Monsieur de Monts' party lost thirty-six out of eighty to *mal de terre*—"land disease." In a mix of wild speculation and accurate guesswork, the expedition's chronicler, Monsieur Lescarbot, attributed the disease to the "great rottenness in the woods during the rains of Autumn and Winter," to "rude, gross, cold and melancholy meats," and—cutting through Lescarbot's rhetoric—to unrelieved lust.

Only his citing of the "melancholy meats" was close to the mark, for scurvy is above all a nutrition-deficiency disease. Whatever the weak spots of his analysis, Monsieur Lescarbot did note the curious connection between scurvy and stress—one that's been cited by modern researchers as well. The disease "found out" sailors who were "grudging, repining, never content, idle." As for a cure, Lescarbot's prescription might be called the French school of scurvy remedies: rich butter sauces to be served with capon and fresh game, plus tender springtime herbs, a happy spirit, and plenty of sex, although this last element was allowed only within the confines of a Church-approved union. "It resteth necessary . . . for every one to have the honest company of his lawfull wife: for without that, ones minde is always upon that which one loves and desireth; the body becomes full of ill humours, and so the sickness doth breed."

With some exceptions—such as Sir Francis Drake's circumnavigation of the globe in the 1570s in the *Golden Hind*, when his crew "refreshed" themselves with South America's wild herbs, the crayfish of the Celebes, and West Africa's lemons and fresh oysters—the British fared no better on long voyages than the Portuguese, Spanish, and French had before them.[2] The Plymouth colony itself lost half of its original group of English settlers to scurvy, while the colony's later settlers treated themselves with lemon juice brought by ship. "This is a wonderful secret of the power and wisdom of God, that hath hidden so great and unknown virtue in this fruit," wrote Sir Richard Hawkins of the British navy after his scurvy-ridden crew had been saved by lemons and oranges upon landing in southern Brazil in 1593. Despite the strength of many testimonials like Hawkins' for the curative powers of citrus fruits, and his observation that in twenty years at sea he'd seen probably ten thousand men consumed by scurvy, English authorities went on debating the efficacy of treatments such as drinking seawater, gargling with sulfuric acid, and—here's the British school of scurvy cures—chugging pints of wort made from soaking malted barley in water. The theory was that

---

2. A British seaman's typical daily diet in 1740 consisted of the following: 1 pound of biscuit, 1 gallon of beer or a ration of liquor, 1 pound of salt beef or pork, 2 ounces of cheese, 1 ounce of butter, 7 ounces of oatmeal, and 4.5 ounces of dried beans or peas. From *The History of Scurvy and Vitamin C* by Kenneth J. Carpenter (Cambridge: Cambridge University Press, 1986).

the wort would ferment inside the patient as it did inside the beer-brewing vats back in England and work some unknown therapeutic effect.

It wasn't for another two centuries after Hawkins' voyage and testimonial that Gilbert Blane, a socially prominent physician who had served with the British fleet in the West Indies and treated dozens of scurvy cases, finally persuaded the Office for Sick and Hurt Seamen and the Lords of the Admiralty to approve the now-famous daily issue of .75 ounce of lemon juice per man. The edict went into effect in 1795; over the next twenty years, the British navy issued 1.6 million gallons of lemon juice (later changed to lime juice, although the terms back then were often used interchangeably). The allowance dramatically cut the incidence of scurvy, helped the British navy defeat the French and Spanish fleets at the Battle of Trafalgar in 1805, and dubbed British citizens at outposts around the globe with the name "limeys." "The British Empire," one medical authority has written, "blossomed from the seeds of citrus fruits."

Scurvy, however, by no means had been eradicated. It still cropped up during Europe's Great Potato Famine and the California Gold Rush of the late 1840s, during polar expeditions, and during the American Civil War when it caused a quarter of the twelve thousand deaths at the Confederacy's infamous Andersonville Prison. In the late 1800s, doctors in both Europe and the United States saw growing numbers of cases of "infantile scurvy" as babies' diets shifted to more processed foods and cow's milk and away from highly nutritious breast milk.

Though it was now relatively easy to cure, no one knew exactly what caused scurvy. A big breakthrough occurred in 1907 when two Norwegian scientists attempted to induce beriberi—another nutrition-deficiency disease afflicting Norwegian sailors at the time—in guinea pigs by feeding them nothing but bread. Instead they induced scurvy, discovering that guinea pigs are, like humans, subject to the disease. The guinea pig discovery made possible widespread scurvy experiments on animal subjects, and research progressed rapidly. Four years later, in 1911, a young Polish chemist by the name of Casimir Funk floated a radical theory that scurvy and several other diseases resulted from diets that were each deficient in different substances that he called "vital amines." Other researchers worked with rats and searched for unidentified dietary substances they called "fat-soluble factor A" and "water-soluble factor B"; scientists hunting the active element in foods that prevented scurvy logically called it "accessory food factor C." Though these unknown substances weren't necessarily amines—a type of base compound—Casimir Funk's term "vital amines" was soon adopted in shortened form. We thus take our daily allowance of vitamins A, B-complex, and C instead of, as was apparently proposed, a dose of "funkians."

Credit for the actual isolation of vitamin C finally went to a scientist from Hungary, biochemist Albert Szent-Györgyi, who was working on an altogether different chemical problem when he isolated a crystalline substance from the adrenal glands of oxen and later from cabbage, oranges, and paprika. The crystals' molecular composition resembled that of common sugars such as glucose and fructose, known chemically as the hexoses, and, unsure of his crystals' exact molecular structure, Szent-Györgyi wanted to call it "ignose" or "godnose." The editor of a scientific journal made him change it to "hexuronic acid"—later known as ascorbic acid or vitamin C. Although he hadn't set out to find vitamin C, his work won him the Nobel prize for physiology and medicine in 1937. "The whole problem," he wrote thirty years later, "was, for me too glamourous, and vitamins were, to my mind, theoretically uninteresting. 'Vitamin' means that one has to eat it. What one has to eat is the first concern of the chef, not the scientist."

Phil swayed slightly from the Bordeaux and the wind-driven rock of the *Myrna Lay* as he stood in the salon and gazed into the open food lockers. He'd skipped the oranges at the supermarket—too much trouble to peel. He'd passed up the apples—too mushy. He hadn't even considered the lettuce or red and green peppers because he preferred to avoid all the mouth-crunching effort a salad demanded of its eater. No, he preferred concentrated foods. And, like the sailors of two and three centuries ago, he had them aboard—in abundance. *Lack* of food wasn't the problem. It was *quality* of food. He rummaged around behind the bags of chips and canned ravioli and diet cola and gummi bears. He found a tin of processed cheese, a stick of plastic-wrapped garlic salami, a box of saltine crackers.

This wasn't going to do—not at all. He unconsciously let go of the locker door and moved his right hand to his upper incisors. He tried to wiggle them with his fingers. Already they felt a little loose. The thought of eating the sticky gummi bears with scurvy-loosened teeth made him want to heave the bag of candy overboard. Or was he just imagining the teeth? He was only three days' sail out of Seattle, after all, not three months'. But maybe his body's stores of vitamin C had been well depleted before he left; scurvy might strike him far sooner than the usual two or three months. He did feel a little weak. That was the first sign. Weakness could set in days or weeks before the loose teeth and other really gruesome symptoms appeared. How low had his body's stores of vitamin C been before he embarked? That was the key question. *Really* low, probably. He tried to remember the last time he had eaten a salad. Or a piece of fruit. Or any green vegetable at all. He wondered if he could count the green olives on the double-mozzarella pizza he'd stuffed into his face while arguing with Myrna during their final date.

Phil plunged his hands back into the upper locker, digging for something to eat that might contain vitamin C.

It's not clear exactly how much Steller knew about scurvy cures when the *St. Peter* put in at the island off the Alaska Peninsula. He was familiar with the scurvy grass—*Cochlearia*—that some European physicians and navigators had been prescribing to treat the disease for more than a century and possibly much earlier.[3] With his insatiable botanical curiosity, Steller apparently had also noted what plants were eaten for their health by the natives of Kamchatka, where he'd spent several months prior to the expedition's departure. Although he was brusquely denied help at the island stop, Steller nevertheless went ashore on the island and gathered what medicinal plants he could. In addition to a native variety of scurvy grass, these plants included a species of dock (a member of the buckwheat family) that was eaten raw by Siberian Eskimos, as well as gentian and "other cresslike plants that I gathered only for my use and the Captain-Commander's."

As Steller gathered plants, one of the sickest sailors, Nikita Shumagin, was brought ashore by the water carriers in the hope that simply being on land would do him good. (One scurvy theory of the time had it that humans needed exposure to certain mysterious "earthy

---

3. William Shakespeare's son-in-law, the Stratford-on-Avon physician John Hall, left detailed notes of successfully treating on-land scurvy patients with scurvy grass. These included his wife—Shakespeare's eldest daughter—the forty-seven-year-old Susanna, whose nasty 1630 scurvy case was accompanied by "Pain of the Loins, Corruption of the Gums, Stinking Breath, Melancholy, Wind, Cardiac Passion, Laziness, difficulty of breathing, fear of the Mother, binding of the Belly, and torment there, and all of a long continuance, with restlessness and weakness." Quoted in *The History of Scurvy and Vitamin C* by Kenneth J. Carpenter (Cambridge: Cambridge University Press, 1986).

particles.") Soon after he arrived ashore, Shumagin died. Named by the men of the *St. Peter*, the Shumagin Islands off the Alaskan Peninsula still commemorate their first fallen crewmate.

Predictably, the other ill sailors in the crew of the *St. Peter* at first wouldn't touch the plants Steller had gathered on the Shumagin Islands. But their tune soon changed. "Ungrateful and coarse men though they were, my ministrations, under divine grace, very clearly caught their attention when, simply by giving him raw scurvy grass, I managed to bring the Captain-Commander—so bedridden with scurvy that he had already lost the use of his limbs—so far within eight days that he was able to get out of bed and on deck and to feel as vigorous as he had been at the beginning of the voyage. Likewise, the *Lapathum* [dock] I prescribed to be eaten raw for three days firmed up again the teeth of most seamen."

Apparently the supply didn't last—through their own neglect at refusing to help Steller gather it. The ship weighed anchor on September 6 and inched its way again across the North Pacific into an onslaught of autumn storms that greatly hindered its progress and allowed the scurvy to take hold again. The wind screamed and whistled through the rigging. Waves battered the *St. Peter*'s hull like the concussion of a cannon shot. The veteran steersman of fifty years' experience, Andreas Hesselberg, had never witnessed storms so violent. "Every moment we expected the shattering of our ship and no one could sit, lie down, or stand," Steller recorded. "No one could remain at his station, but we were drifting under God's terrible power wherever the enraged heavens wanted to take us. Half our men lay sick and weak, and the other half was healthy out of necessity but thoroughly crazed and maddened by the terrifying movement of the sea and ship."

On October 2, during a lull between storms, Steller took stock: twenty-four men sick and two dead. The gales had blown the ship backward toward America some 230 miles. The supply of brandy ran out. But the storms continued, and so did the scurvy. By October 20, a man was dying nearly every day. Only ten men, including Steller—who pointed out that it wasn't in his job description but who helped sail the ship nonetheless—had enough strength to work the *St. Peter* at all. Even they were so weak that the *St. Peter* didn't put in for fresh water at a small island they spotted because the decimated crew couldn't have hauled up the anchor from the bottom.

It was then that the two highest-ranking officers below the ailing Bering, Master Khitrov and Lieutenant Waxell, issued a strange and sudden order—strange and sudden, according to Steller, anyway—to change course and head more northward, thus "criminally deviating" from the agreed-on plan to aim straight for the *St. Peter's* home harbor at Avacha on the Kamchatka Peninsula. Soon after, the officers announced with "mathematical certainty" that the *St. Peter* was about to make landfall on the Kamchatka Peninsula. The crew, at least those who could still walk, gathered on the *St. Peter's* deck early on the morning of November 5, 1741, to witness the truth of this prediction. "To the astonishment of all," reported Steller, "it came to pass that at nine o'clock land was sighted."

After joyous celebration and prayers of thanks, the officers and men examined sketches of Avacha and compared them to the features of the coast ahead of them, and the two were found to agree completely. They even thought they spotted the mouth of Avacha's harbor and its lighthouse. The *St. Peter* sailed in closer. The sun appeared, and the officers took a noon bearing with the sextant. But something was amiss. The observation located them between 55 and 56 degrees north latitude instead of down near 52 degrees, where Avacha lay and where they believed they were from the look of the coast. They attempted to work the ship back to the south, near to the point they'd identified as Cape Isopa, but strong headwinds hindered them. Another gale rose, and the crew no longer possessed the strength to climb aloft and furl the sails. During the night, the raging wind and the unfurled sails whipping about shredded the ropes that supported the mainmast.

The next morning, November 6, the ship's officers and commander argued about how to proceed. Bering insisted they keep sailing south until they reached the certain safe haven of Avacha, although the mainmast was now too weak to carry much sail and the crew even weaker. However, Master Khitrov—"professing if this were not Kamchatka he would let his head be chopped off"—went about convincing the sailors that the *St. Peter* should put into land immediately on the unknown coast. The crew sided with Master Khitrov, and when one lower officer objected to Khitrov's plan, Khitrov and Waxell tossed him out of the meeting. "Out!" they shouted, according to Steller. "Shut up you dog, you son of a bitch!"

That night, as the *St. Peter* put into a cove of the unknown shore, high seas suddenly rolled in, tossing the ship up and down "like a ball." The superstitious among the sailors unceremoniously pitched the recent dead, who had been saved for a land burial, head over heels overboard in hopes of calming the tempestuous sea. The next morning Steller—still healthy, presumably due to his regimen of raw herbs, and no doubt eager to get off the ship—was the first ashore along with the surveyor Pleniser, Steller's cossack servant, and a few of the sick. He carried with him as much of his luggage as he could; given the poor anchorage and lack of healthy crew, he knew that the first storm would blow the *St. Peter* out to sea or smash her on the beach. He spent that first day, November 7, exploring, noting particularly how little fear of humans the sea otters showed—an animal that was heavily hunted on the mainland for its pelt—in the way they scampered down the beach and into the water to greet their rowboat as it landed. Later, returning to the landing spot, he found Lieutenant Waxell, weak with scurvy, taking the air on shore.

"God knows if this is Kamchatka," Steller said.

"What else would it be?" Waxell replied. "We will soon send for posthorses."

Over the next two days, Steller and his two companions—his cossack hunter and servant, Thoma, and his friend and fellow German, the surveyor Pleniser—further explored the unknown shore. Steller found more clues—blue foxes that were extraordinarily cheeky, the total absence of trees or shrubs on the mountainous land, the "sea cows" lying on the beach that Thoma said he'd never seen before on Kamchatcka, and the presence of sea clouds in the distance that indicated water all around.[4]

Perhaps, Steller thought, they'd beached on a large island.

No doubt the trio mulled over this prospect—or rather, Steller expounded on it to the other two—as they sat around their campfire the first night on shore. As they sat there, Steller reported, a blue fox strolled into their camp and before their eyes walked out with two of the ptarmigan—a type of grouse—they'd hunted. In what proved to be a pivotal moment of the voyage, Thoma, who was also suffering from scurvy, suddenly turned to Steller and angrily blamed Steller and his curiosity for all of his troubles.

Steller reported his reply: " 'Cheer up!' I said. 'God will help. Even if this is not our country, we still have hope of getting there. You will not die of hunger. If you cannot work and wait on me, I will wait on you. I know your honest heart and what you have done for me. Everything I have is yours also. Just ask and I will share with you half of all I have until God helps.' "

The conversation started Steller thinking about preparations for winter, should the unknown shore turn out to be an island. At his urging, Pleniser agreed to help him build a hut, and they entered a pact to help each other as friends. With makeshift shovels the threesome soon began to excavate a pit house with a roof supported by driftwood similar to those built for winter dwellings by Kamchadal natives. Three other German crew members of the *St. Peter* joined

---

4. Also called Steller's sea cow, this marine mammal grew as long as 25 feet, propelling itself through the water with a fluked tail to feed on seaweed. It was hunted to extinction for food by Russian sealers within thirty years of the time it was first described by Steller, who also first described hundreds of other plants and animals in Siberia and America and for whom the Steller's jay and Steller's sea eagle are also named.

their household, plus two more cossacks, and Bering's two servants, all of them pledging to "enter into a community of goods."

Within a day or two, the other members of the *St. Peter*'s crew who still had strength enough to row to shore began digging out their own underground huts in imitation of Steller and organizing their own households. Steller, no longer the scorned physician and botanist, had suddenly become by example a leader. Master Khitrov's star, meanwhile, was sinking fast. The ill sailors blamed him for their misery and isolation. He begged and wheedled the Germans for a corner in their hut so he wouldn't have to sleep with the ordinary seamen, who were uttering threats against him day and night. Steller and his housemates, however, were "offended by him, and we refused him all hope, because he was healthy and lazy, and he alone had plunged us into this misfortune."

Work continued on the huts on November 12, a week after land had first been sighted, and was completed the following day. But their shelter proved too late for many of the crew.

"That day many sick were brought from the ship," Steller wrote, "among them some who, like the cannoneer, died as soon as they came into the air, others in the boat on the crossing over like the soldier Savin Stepanov, others once on shore like the sailor Sylvester. . . . Even before they could be buried, the dead were mutilated by foxes that sniffed at and even dared to attack the sick—still alive and helpless—who were lying on the beach everywhere without cover under the open sky. One screamed because he was cold, another from hunger and thirst, as the mouths of many were in such a wretched state from scurvy that they could not eat anything on account of the great pain because the gums were swollen up like a sponge, brown-black and grown high over the teeth and covering them. . . . The barracks was finished, and during the afternoon we carried in a lot of the sick. But because of the narrowness of the space, they were lying about everywhere on the ground, covered with rags and clothes."

Though he did not give a precise description of the men dying on the beach, Steller probably saw all the classic symptoms of scurvy. Typically, the sailor would first start to weaken before any outward physical marks appeared. The first of these was small purple spots on the sailor's buttocks and the back of legs as small hemorrhages formed around the hair follicles. His skin turned dry and rough. He sometimes couldn't catch his breath. His joints ached deeply, especially his knees, hips, and ankles, because his weight alone caused trauma inside the joints that led to swelling. His legs swelled and hurt from hemorrhaging within the muscles and tissues beneath the skin. He bruised with the slightest touch. Old scars reopened as the tissues pulled apart. Scratches and wounds didn't heal. He became so weak he couldn't lift his own limbs. As their capillaries ruptured, those most famous symptoms of scurvy appeared—his gums swelled into black and purplish masses that were too tender to chew food and which bled with the slightest scrape. His teeth loosened in his skull. If he'd had any strength at all, he could have plucked out his own teeth with his fingers alone.

"If the situation continues," as one scurvy description puts it, "the body will degenerate into a bleeding pulp for which death is a blessing."

What finally happened to the sailor—moaning on the cold beach in pain and cursing his officers and his gods for bringing him there—was that he overexerted himself by, for instance, trying to crawl into the shelter of the newly built underground barracks. This sudden exertion burst scurvy-weakened major blood vessels near his heart. He dropped dead, facedown into the frozen sand.

Phil had now torn most of the food packages out of the upper and lower lockers and flung them to the salon floor in his effort to find something fresh, something that might contain some vitamin C. Rolling out on the floor went the cans of chili and refried beans. Between the crashing of the cans he could hear the rain thumping on the deck overhead with renewed vigor. How long might he be pinned in this cove? Down went a box of taco shells, a can of anchovies, an extra tube of toothpaste. At least he'd remembered his toothbrush; he hadn't totally neglected his health. But last night hadn't he spotted a streak of blood when he spit out the toothpaste into the little stainless-steel sink in the *Myrna Lay*'s head? Or was that just the detritus from a red gummi bear? Or the dregs of the bottle of Haut-Médoc? He imagined the gums that belonged to Bering's sailors dying on the beach. How long would it be before his own gums looked like that? The thought of scraping the extra-stiff bristles of his toothbrush over those pulpy, purplish black masses of dying tissue—the brushing action tearing off strips of flesh—made him want to gag.

Phil momentarily abandoned his search to run to the head and check his mouth in the mirror.

There is a brief but fascinating history of researchers who have induced scurvy in themselves, intentionally or not. It started in the mid-eighteenth century when one William Stark—a young, eccentric English physician who'd met Benjamin Franklin and who may have taken a little too seriously the ever-practical Franklin's commendation of a simple diet—managed to kill himself by eating mostly bread and water.

In 1939, a surgeon at Harvard Medical School, John Crandon, intentionally induced scurvy in himself, developing the first signs of it after twelve weeks on a diet that, in the eyes of some undergraduates, no doubt contains all the key food groups—bread, crackers, cheese, eggs, sugar, chocolate, and beer. A few years later, during

World War II, a small group of conscientious objectors in Sheffield, England, volunteered for an experiment in which the vitamin C had been boiled, aerated, or otherwise destroyed in milk, potatoes, and other foods they were fed, after the volunteers had initially boosted their vitamin C levels with large doses. After four to eleven weeks on the diet, however, the vitamin C level in their blood plasma had dropped to almost undetectable levels. By twenty weeks, all of them showed little plugs forming in the hair follicles of their buttocks and backs of their legs, and six of ten had hemorrhages beneath the skin. After the twenty-sixth week, wounds that had been intentionally inflicted on the volunteers—who by this point might have wished they'd chosen the war front instead—no longer healed.

By the thirty-sixth week, the gums of nine of ten had become "purplish, swollen and spongy" and "special incidents" occurred. One of the volunteers, a twenty-two-year-old student, after heavy exercise suffered what was diagnosed as a heart attack. Another awoke in the night with chest pains and a systolic murmur. A chest X ray in another had revealed the sudden growth of a small, preexisting tubercular lesion. Immediate, massive injections of vitamin C cured all the special problems. Researchers then reined in the experiment, giving each of the remaining volunteers 10 milligrams of vitamin C each day—only about one-sixth of today's recommended 60-milligram daily intake in the United States for healthy adults and about one-fourth of the 25- to 40-milligram daily intake recommended by Canada. Even with this minimal dose of vitamin C, within two weeks the volunteers had showed marked improvements and after ten to fourteen weeks had fully recovered.

Using sophisticated biochemical analysis, researchers have since understood that vitamin C serves as a kind of atomic welder in the body's protein-making factories. Humans constantly produce protein in order to grow bigger or to replace worn-out parts—a half a ton of it during a child's first few years alone and over a lifetime some five tons of it, which would halfway fill a dump truck. This is forged inside the cells' tiny ribosomes, which select an array of the body's twenty amino acids and fasten them together in different combinations. One of the most important of these proteins is collagen—an extremely tough but flexible material that forms the basis of the body's connective tissues such as ligaments and tendons as well as bones, and forms layers within blood vessel walls and skin. (Leather, for instance, is preserved animal collagen.) In essence, vitamin C, or ascorbic acid, tacks an extra group of oxygen and hydrogen atoms onto some of the amino acids that form collagen, causing them to bond extremely tightly together and giving the collagen its tough, flexible quality. It plays a similar but less clear role in the production of elastin, a protein in the body's elastic fibers, including those in veins and arteries.

As the crew of the *St. Peter* lay moaning on the beach, their tissues and blood vessels containing these proteins, without the bonding influence of vitamin C, were literally coming unraveled like the shredded rigging aboard the *St. Peter*. Captain-Commander Bering was among the very sick. He had been carried to shore and placed in a sailcloth tent on the sand, whereupon he asked Steller what he thought of this new land.

"It does not look like Kamchatka to me," Steller said.

"The ship probably cannot be saved," Bering replied. "May God spare us our longboat!"

Steller frequently checked in on Bering; the captain's two servants now mostly looked after their own survival. He gave Bering drinks of water and brought food, such as a baby sea otter still nursing on its mother that Steller knew contained plenty of nourishment. "But he declared a very great aversion to it," wrote Steller, "and wondered at my taste, which was adapted to the circumstances and place."

Once again the officers of the *St. Peter*, to their great detriment, had shown rigidity where Steller showed flexibility and improvisation. Though he "refreshed" himself with ptarmigan, Bering sunk deeper into illness and despair. His body teemed with lice. He had himself buried in the sand from the waist down.

"The deeper in the ground I lie," he said, "the warmer I am."

Two hours before daybreak on December 8, 1741, Captain-Commander Vitus Bering died. "He would undoubtedly still be alive," Steller wrote, "if we had reached Kamchatka and he could have had the benefit of a warm room and fresh food. As it was, he died more from hunger, cold, thirst, vermin, and grief than from a disease." Still, Bering's body showed abundant signs of scurvy: swelling of the feet and joints, as well as decay—"mortification"—of the tissues. Composed and even "blissful" as he died, Bering retained his reason and speech until the end. "He wished nothing more than our deliverance from this land," Steller wrote, "and from the bottom of his heart, his own complete deliverance from this misery. He might well not have found a better place to prepare himself for eternity than this deathbed under the open sky."

Phil stared at his gums in the head's mirror over the little sink, illuminated by the 12-volt lightbulb. They looked normal—kind of. Maybe they appeared a bit darker than usual. And his teeth had a reddish tinge. Was that the Bordeaux, or the first sign of hemorrhaging capillaries? But purple spots on the backs of his legs and buttocks should have showed up first. Phil quickly unzipped his jeans and peeled them and his boxer shorts down to his knees. He pulled up his shirttails and twisted his head around. He couldn't see. He ran his right hand over the skin of his buttocks. He felt small bumps. *Damn!* He twisted farther. On the back of his thighs and his buttocks *there were at least four or five small purple spots!* Were they just pimples? Or was it the onset of perifollicular hemorrhage?

Pulling up his pants with one hand, Phil hurried back into the salon, kicking with his bare feet through the dozens of scattered packages and cans of food. There had to be some vitamin C *somewhere* in this mess.

Bering's death marked the end of the rapid die-off of the crew. By Christmas, the survivors had recovered their health if not their full strength. A storm, meanwhile, drove the *St. Peter* ashore and buried her hull in sand, and she was deemed no longer seaworthy. Search parties returned to report that they were, as Steller had predicted, on a large island—one that showed absolutely no signs of human habitation. Whatever attempts the marooned party would make at rescuing itself would have to wait for spring.

Hunting sea mammals for their basic food, the castaways settled in for the winter. Like the Eskimos, who traditionally have lived scurvy-free, the sailors found that an all-meat diet could provide them with the nutrients they needed—especially the organ meat, eaten raw or lightly cooked. A 3.5-ounce chunk of raw seal liver alone can contain enough vitamin C to meet the Canadian daily recommendation, as do two or three pounds—not an unusual amount on an all-meat diet—of lightly cooked seal flesh, as exposure to oxygen and excessive heat destroys the vitamin. The Yukon Indians ate the raw, vitamin C–rich adrenal glands of field mice to cure scurvy. Even today, native hunters in Manchuria, Greenland, and other northern climates extract the still-warm kidneys and liver of deer, seal, and other animals they have just killed and hungrily swallow them on the spot like giant, raw, dripping capsules of vitamin C. They also love to chew on vitamin C–laden whaleskin. These mammals—like all other mammals—need vitamin C as badly as humans but have their own mechanisms for manufacturing it from simple glucose. The only known land mammals that cannot make their own vitamin C are the guinea pig, the fruit bat, primates (including humans), and a bird called the red-vented bulbul—all lacking a key enzyme for the final conversion from glucose to ascorbic acid.

With the fresh meat, the ascorbic acid level in the sailors' bloodstream began to rise, and their organs absorbed it. Their adrenal glands took in large amounts in order to manufacture norepinephrine and epinephrine, which are secreted when the body is under stress to kick up the heart rate, dilate blood vessels, and release extra fuels into the bloodstream to prepare for fight or flight. Thus people under stress—like the sailors on the Bering expedition—may need more vitamin C and are more susceptible to scurvy.

Assuming they were consuming at least 10 milligrams of it per day in their sea-mammal diet, the vitamin C would have started to restore the crew's health within two weeks. Their strength returned, their teeth firmed up. Within eight weeks their skin would have resumed its normal coloring, except for brownish patches where the hemorrhages had occurred. After ten to fourteen weeks their gums would have totally recovered, judging by the results of the World War II experiments with the conscientious objectors. Still, on the island complete recovery may have been slower.

During the long winter, the party suffered a severe outbreak of gambling accompanied by quarrels and strife. All gambling was finally banned by the petty officers. In spring, work commenced on the construction of a new, smaller boat out of the salvaged timbers of the *St. Peter*. From chaos—a "state of nature," as Steller called it—the renourished survivors organized themselves into a functioning society. Everyone signed a pact: the twelve men who knew how to wield axes would work full time as carpenters on the new boat, and the remainder would divide into two hunting shifts to ensure a constant supply of meat. Under this common purpose—their deliverance from the island—work proceeded quickly as the snow melted. Their emaciated bodies were nourished by runs of fish so abundant that a single haul of their old nets caught enough to last them eight days. Aided by Steller's keen botanical eye, they savored Kamchatkan sweet grass, bulbs of sarana lily, roots of wild celery, oyster-leaf, shoots of fireweed, and salads of scurvy grass, brooklime, and bitter cress, all washed down with tealike infusions of lingonberry and wintergreen. Lieutenant Waxell reported that he didn't recover his full strength until eating these greens with the coming of spring.

By mid-July, they'd built the new hull. For the next month they fashioned the rigging, spars, and mast, forged the hardware, baked ship's biscuit from their salvaged flour supplies, and put up barrels of salted sea cow. On August 13, the castaways boarded the 42-foot-long boat, a one-masted hooker also named the *St. Peter*. Out of the original crew of 78, only 46 had survived. There was nevertheless barely room to jam themselves in to sleep between the supplies in the hold and the deck planking—the supplies including nine hundred valuable otter pelts they'd hunted on the island. A fair wind and sunny skies propelled them past the southern tip of Bering Island, as they christened it. They looked one last time at the mountains and valleys they'd come to know so well. "God's grace and mercy became manifested to all," wrote Steller, "the more brightly considering how miserably we had arrived there on November 6, had miraculously nourished ourselves on this barren island, and with amazing labor had become ever more healthy, hardened, and strengthened; and the more we gazed at the island on our farewell, the clearer appeared to us, as in a mirror, God's wonderful and loving guidance."

Barefoot in the salon, his pants hitched halfway up his hips, Phil plucked from the floor the most nutritious-sounding items and, holding the packages toward the light of the brass lamp, read the nutritional labels printed on the back.

A can of chili—zero vitamin C.

He tossed it back to the floor and snagged the can of refried beans. Zero.

Down it went again. There was the toothpaste—*no way*—and

the sugar and coffee—*forget it*. A box of macaroni and cheese, a bag of powdered-sugar doughnuts, a package of spaghetti. All o percent.

Phil then inadvertently stepped on a bag of barbeque potato chips. It made a loud crunch. He wasn't even going to bother to pick it up. But if the Great Potato Famine had triggered a widespread epidemic of scurvy, potatoes must contain some vitamin C that had kept it at bay.

He reached down. He held the bag's silvery back to the lamplight. He couldn't believe it. A single serving—about fifteen chips—contained 10 percent of the 60 milligrams that was the U.S. recommended value of vitamin C, in other words about 6 milligrams. If the conscientious objectors during World War II recovered from their experimental scurvy on a mere 10 milligrams daily, this meant that two big handfuls of potato chips daily would be enough to stave off or eventually recover from a case of scurvy. If the Bering expedition had only carried enough potato chips, Phil realized, they would have survived just fine.

Phil ripped open the bag and stuffed a fistful of the spicy, salty chips into his mouth, chewing them as he kicked through the remaining cans and packages and jars, the orangish crumbs falling from his lips. A jar of salsa bumped his toe. That was about as close to a fresh vegetable as anything he'd stowed aboard the *Myrna Lay*. He checked the label. *Another score!* Two tablespoons of the salsa—one serving—contained 8 percent of the daily RV. It must be in the vitamin-rich peppers. He twisted open the lid and jammed another fistful of chips into it, pushing the dripping chips into his mouth. The salsa trickled down his chin and splattered to the floor.

He looked down. At his feet lay a package of grape Kool-Aid. He picked it up and read the label. Each serving contained 10 percent of the daily value of vitamin C. He dug through the galley cupboards, and found a plastic pitcher and mixed up a quart of the Kool-Aid. He took a long swig straight from the pitcher. It was delicious. He could almost feel the vitamin C flowing out through his limbs like the sap that rises in a tree with the spring sun. Those peri-follicular hemorrhages—or pimples, whichever they were—wouldn't stand a chance.

It was then, on the little shelf above the galley sink, that he spotted the package wrapped in brown paper Myrna had left on board along with the Steller journal. In his anger at her, he'd stashed the package there before he'd cast off. Now it occurred to him that maybe it contained food. Maybe even fresh fruit.

He snatched the package off the shelf and tore it open. Inside was a small cardboard box. He ripped apart the box. There was tissue paper stuffed inside it. He could hear something rattling inside the tissue paper. He shredded the tissue and found a small plastic bottle. He held it up to the lamplight. The plastic was dark-colored—strong light could destroy the potency of its contents—and the label was marked "Vitamin C Supplement, 500 mg."

Taped to the bottle was a note. Phil read it. "To keep you safe. Love, Myrna."

Smiling, he opened the bottle, tossed two of the pills into his mouth, and washed them down with Kool-Aid.

CHAPTER SIX

WARM TO

THE TOUCH:

HEATSTROKE

The top of the pass and she's won. The rest will be downhill, fast, the wind whistling through her helmet and whirring through her spokes. They'll never catch her then. The top of the pass and the prize money is hers. Hers and her family's—her mother and father, younger sister and brother, her wild-haired, curse-spewing grandmother, still shouting at imaginary Nazi invaders fifty years after they left—a family who shares an apartment in a crumbling concrete building in an Eastern European city. The top of the pass and she'll escape this brutal heat, dive down the far side into a cool breeze, into a stream of those green American dollars.

She must be close. From the early-morning start at the chamber of commerce, which sponsored the race hoping to spur the region's economic development, the course funneled the competitors along a strip of fast-food restaurants and car lots, and then up, up, up the winding, narrow road into the Appalachian Mountains. She made her breakaway on the first big hill. Her coach has always said, "Attack on the climbs, defend on the downhills. It's on the climbs where races are won."

She looks back. No one in sight.

She's always been a good climber but has never made a breakaway half so bold as this. Back there, somewhere, is the rest of the pack, and somewhere in the pack are the three best hill climbers in the world—the Italian girl and the French girl and the Colombian girl, famous riders all. She is a nothing to them, a nobody. She knows it, and they know it, but this is exactly what she plans to work to her advantage. For now, they let her go sprinting away by herself in front of the pack, figuring they could easily reel her in once she'd exhausted herself. She'll surprise them. That's her strategy: to get so far ahead on this climb that they never catch her.

How many more switchbacks on the road? she wonders. It can't be far. It's so hot. Her thighs burn with the strain of pumping uphill, and her neck and arms and shoulders and back join in the pain, the muscle groups working in tandem to shove her entire body down like a piston to power each stroke of the pedals. Did she break away too soon? If she'd waited much longer, they wouldn't have let her get away. It's so hot. Short, quick breaths exhaust rather than revive her—like inhaling steam. *Breathe deeply,* she reminds herself. It's so hot that the pace car overheated on the first steep hill, spewing a column of steam beside the road as the pack whisked past. Then she stood on her pedals and sprinted away from the others as they rode in a stately bunch, and she was in front, by herself, with only open road ahead that wound ever upward into the leafy hardwood forest.

Push, push, push, she tells herself.

It's so hot. The sweat rolls down her neck, her back, her arms, her legs. It's so hot—it's predicted to reach 95 degrees today, they said at the start—that a bluish haze of moisture saturates the air. The beech trees along the shoulder look as if they were painted by an Impressionist who blurred their green edges and laid in thick splotches of dark oil to render the shadows within the forest. Her coach warned her about the Appalachian humidity—nothing she's ever experienced in Europe. She should have come two weeks early to acclimatize herself, as he suggested, instead of staying with her family, where it was cheaper, and training each day in the cool Baltic air. She should have listened more closely to her coach. Now he's not here, either—there's no longer enough money back home to send him abroad. For this race, her strategy is her own. She's decided that this is her moment to put it all on the line.

To lie down in those dark, thick Impressionist shadows. But first the prize money. Just the top of the pass, and the prize money is hers. The rest is downhill, easy to defend. *Concentrate!* she tells herself. She focuses on the smooth, circular stroke of her pedals and the rhythm of her breathing, ignoring the burning pain in her thighs. Just the top of the pass, that's all she needs, and then the cool breeze, and the money, is hers.

But it's so hot.

Jesus Christ possibly was a heat victim. According to one medical theory, crucifixion, an execution method favored by the kings of ancient Persia and the Mediterranean, relied on the sun's rays to do the job cheaply and effectively. (Another theory attributes death by crucifixion to difficulty working the breathing muscles.) With the victim—often a pirate or political agitator—nailed or tied upright to a wooden cross and exposed to the full heat of the sun, the veins in the legs and other extremities dilated as part of the body's cooling mechanism. But if the legs remained motionless long enough, and there was no muscular contraction to help squeeze the blood from the lower extremities upward, large volumes of blood collected in the legs' swollen veins and arteries, and not enough was pumped up to the brain. For this reason military guards standing at attention in the heat are instructed to contract and release their toes, to keep the leg muscles working. If not, they can suddenly collapse, regaining consciousness once they're stretched out on the ground and blood can flow again easily from the legs to the brain.

Propped upright on a cross for hours without relief, however, the crucifixion victim's state of unconsciousness deepened, his face grew cold and white, heartbeat and breathing faded, and eventually he died, a process that was sometimes hastened by shattering the transgressor's legs with a metal bar, which undoubtedly ruptured veins and arteries and caused further loss of blood. One historical curiosity of this medical condition brought on by crucifixion—known as orthostatic hypotension (upright-standing low blood pressure)—is that the victim can appear dead before actually dying and, laid horizontal in a cool place for a few hours, can recover fully. Some commentators have raised the question whether this is how Jesus, laid out in his tomb after he was removed from the cross, might have "risen from the dead."

An Old Testament mystique surrounds the notion of death by heat, not the least of it the concept of hell as a burning pit. Judaism, Christianity, and Islam all were born amid burning desert sands, where unbearable heat and thirst provided a fitting punishment for sinners—temperatures that meant paradise to the Vikings, who imagined hell as an underworld of cold and dark. References to heatstroke are sprinkled throughout the Bible, such as Manasses who died after harvesting barley in the hot fields. The Arabians called heatstroke "siriasis"—after Sirius, the Dog Star, which in the hot summer months chases through the sky behind the sun. Roman legions that marched into Arabia in 24 B.C. were wiped out by a much more deadly enemy than opposing armies, discovering that heatstroke, according to ancient historians, "attacked the head and caused it to become parched, killing forthwith most of those who were attacked." Early European adventurers returned from the tropics fearing the sunlight itself, believing it contained deadly "actinic rays" that could penetrate the skull. Up until 1940, the British army issued its tropical units spinal pads and headgear to deflect the fatal rays, creating that symbol of Britannia's colonial potency as well as her bodily frailty in the face of tropical disease and heat—the pith helmet.

On average, about 200 people are reported to die in the United States each year from heat-related causes, although the actual number is probably higher; heat-wave summers can raise the number close to 1,000 and even higher if heat is counted as a contributing factor in other deaths. The 1995 Chicago heat wave alone killed an estimated 700 people in a week's time. During the 1977 New York City power blackout and heat wave, deaths jumped from 197—the city's usual daily average for that season—to 237 and climbed again to 298 when the temperature hit 104 degrees.

Heatstroke can strike anyone. However, it especially seeks out the elderly, the overweight, those with chronic illnesses such as cardiovascular disease, and those who have taken certain medications or consumed too much alcohol. People living alone who have few family members or close friends to check up on them, people living on the top floors of buildings (where hot air accumulates), and people who don't have air-conditioning all are more likely victims of heatstroke. A fan can help, but once the temperature tops 90 degrees, a fan does not offer protection.

Classic heatstroke—the slow rise of body temperature among sedentary people in hot environments—accounts for most heat-related deaths. The other heat killer, exertional heatstroke, attacks the young and the healthy. These are the foot soldiers and field laborers and gold miners, marathon runners and bicycle racers, high-school wrestlers and football players. They are the people who drive their bodies— or whose bodies are driven by taskmasters, real or imagined—too hard in hot or humid conditions, and they are especially vulnerable if they have not drunk adequate water or lack proper acclimatization to heat, a process that takes the human body about two weeks. Heatstroke ranks third behind head and neck injuries and cardiac disorders as a cause of death among U.S. high-school athletes. "Probably no greater strain is put on the human body," writes one medical expert on the subject, "than heavy physical exertion in the heat."

Strangely, the human body can tolerate large drops in its internal temperature, when metabolism slows to a crawl (hypothermia), but can survive only very small rises when metabolism shifts into high gear (hyperthermia). Women resist cold better than men, due to their insulating layer of fat, but heat strikes them harder. When it does occur, heat illness can take many forms, a complex continuum of maladies that traces the curve of the body's rising temperature. First is the mildest of the heat illnesses, heat edema, when feet and legs swell due to the dilation of the veins and arteries after long periods of sitting or standing. Then heat syncope—fainting or dizziness— and heat cramps, painful spasms in the muscles of the legs, arms, and abdomen that can be exacerbated by lack of salt or diuretic medications. Heat exhaustion—a term often confused with heatstroke but a less serious condition—results from sweating excessively in the heat and humidity and is marked by body temperatures over 100 degrees but less than 105, as well as headache, dizziness, nausea and vomiting, chills, weakness, rapid heartbeat, and low blood pressure.

Untreated by cooling, rest, and rehydration, heat exhaustion can rapidly develop into the very dangerous condition known as heatstroke, when the core temperature of the human body soars to 105 degrees and above. Survival depends on just how long the victim has remained overheated, and the condition demands immediate and rapid cooling. Some victims who receive prompt treatment have survived very high internal temperatures, such as Willie Jones, dubbed by the press "the Human Torch," whom doctors rapidly cooled with ice water from an estimated body temperature of 120 degrees after he was found unconscious in 1980 in his Atlanta apartment. Another victim survived a body temperature of 113 degrees after he had a minor accident with a police cruiser, swallowed a gram of methamphetamine he was apparently trying to hide, and, chased by police, fled on foot as his core temperature soared to catastrophic levels. Prompt cooling and rehydration at the hospital brought his temperature down to 100.6 degrees within ninety minutes, he recovered from a coma after twenty-six hours, and he was discharged from the hospital in stable condition after five days. Due in part to its many physiological complications—heatstroke can affect virtually every organ in the human body—estimates of mortality from heatstroke vary wildly, from 10 percent to as high as 80 percent, but whatever the exact number, doctors are unanimous in calling heatstroke one of the few true medical emergencies.

Complicated equations exist to describe the body's heat gain, but they reduce to a simple physiological truth: If your body generates more heat than it's capable of dispersing, you will become hotter.[1] And if you don't stop generating heat—seek shade, lie down, cease your muscular contractions—you will eventually overwhelm your internal thermostat. Your body's temperature then will go haywire. Your cellular metabolism will accelerate wildly, generating even more heat. A fairly predictable sequence of events follows: unconsciousness, convulsions, and death, as the body essentially cooks its own flesh from within.

Push . . . push . . . push . . . Push the pace as hard as she can, beyond what the other riders can handle, push until the distance between her and the pack stretches like a rubber band, tighter and tighter, longer and longer, until it snaps and they'll never catch her.

Her muscled thighs pump up and down in a smooth, circular motion. Her breathing keeps a half-time beat to the rhythm of her legs, a counterpoint played by the precise click of the bike chain as she downshifts and it hops to a larger sprocket of her rear hub. More power, less speed. The road steepens as she swings around a switchback, twisting and tunneling upward through the beeches and maples and oaks. The lush Appalachian scenery for her provides only a passing sensation of light and shadow, of hot and cool, as an animal might perceive the forest when moving quickly through

---

1. $S = M - E +/- (R + C) - (+/- W)$, where S = heat storage in the body; M = rate of energy metabolism; E = rate of evaporative loss of heat; $R + C$ = rate of radiant and convective loss, when the sign is positive; W = work rate, where $+W$ represents work energy done on an external system and $-W$ represents work energy absorbed by the body. Values of W are usually small with respect to M. From *Sports Medicine and Physiology*, ed. R. H. Strauss, M.D. (Philadelphia: W. B. Saunders Co., 1979), p. 137.

it. Her whole focus is instead on the rhythms of her legs, the fine-tuning of her gears to subtle shifts in gradient, the huffing depth of her breathing. *Click*—again she downshifts, her legs still whirling steadily, but now the bicycle moves more slowly, her whole being straining to push it up the steep incline.

Trickles of sweat swell to rivulets. Every nine seconds, each of her two million sweat glands—tightly coiled tubes just beneath her skin that would reach 4 feet if stretched out—suddenly contracts and squirts a few drops of moisture through her pores and then recharges, each hour spilling out over half a gallon of sweat, composed of 99.5 percent water plus dissolved salts. She periodically drinks from the water bottle mounted to her bike frame, but the human body can't absorb water from the stomach—no matter how much one pours in it—any more rapidly than a little over a quart an hour. Even while drinking copiously, she's slowly becoming dehydrated.

In chillier weather, her body's extra heat would be wafted away by cool air moving over her skin—the term for this is *convection*—but at an air temperature of over 95 degrees, sweating is the body's only means of cooling itself. Simply sitting at rest, she would grow hotter, due to her own metabolism, by 2 degrees hourly without some means of cooling down. (The human brain alone, awake or asleep, sharp or dull, puts out as much heat as a 15- to 20-watt lightbulb.) Straining up the switchbacks, she'd reach heatstroke temperatures within 10 to 12 minutes without the sweat that evaporates from her skin surface and carries off much of her body's excess heat. That's assuming, of course, that her sweat evaporates easily, which on this day is a problem.

She pedals upward. Her core temperature still lies within the "fever of exercise" zone—the elevated temperatures that the human body can reach without damage under long and heavy exertion. In a trained athlete, the zone extends from about 100 to 104 degrees, as first measured in runners finishing the 1903 Boston Marathon. There is a very fine line, however, where "fever of exercise" ends and heatstroke begins.

With each powerful downstroke her temperature climbs a tiny fraction of a degree and her physiology adjusts to cool it. Her entire body now operates just like the radiator of a car. She feels a tightness around her wrist as her watchband cuts into her flesh. Just beneath her skin surface her blood vessels swell to carry the hot blood from her contracting muscles and overheated organs to her sweat-cooled exterior. The blood flow to her fingertips alone now is one hundred times what it would be in cold weather. In the heat, the capillaries of her entire skin surface can expand so far that to fill them completely would require a gushing flow of up to eight liters of blood per minute. Her heart, meanwhile, must pump furiously in its attempt to keep them running full.

The road switchbacks left, then right up the mountainside, growing steeper as it makes each bend. *Click* go her gears as she downshifts further, into her lowest gear. It's still not low enough. Her thighs burn. Her pedal strokes drop—from 80 to 60 to 50 revolutions per minute. *Keep it going,* she tells herself, *keep it going.* She stands up on her pedals to throw her full body weight into each downstroke. Heat shimmers off the pavement as her bike crawls up the mountainside, shiny puddles of heat. She longs to hear the zipping splash of her tires through the puddles, the flicking drops of road moisture tossed from the tires into her face, but it never comes.

Instead, sweat rolls down off her body and drops in dark, spidery splashes on the cracked, dry asphalt. When the atmosphere's relative humidity climbs over 75 percent and into the 90th percentile, very little sweat can evaporate into the moisture-laden air, especially without a breeze that whisks a large volume of air past the skin surface. This lack of evaporation was what killed three American high-school and college wrestlers who, in three separate incidents in the fall of 1997, wore water-impermeable suits during heavy workouts in an attempt to sweat off pounds rapidly so that they could compete in lower weight classes. Likewise, once she shifts into her lowest gear and is barely moving up the mountainside, the breeze generated by her forward motion has all but ceased. Her sweat glands are firing wildly, but her cooling system has been rendered nearly useless by the high humidity and heat and lack of moving air.

She feels enveloped in heat, inside and out. Even the patches of shade on the black asphalt are hot. If she had followed her coach's advice and come to the sweltering Appalachians two weeks ago, her body would have learned to dilate its blood vessels and kick its sweat glands into action at a lower core temperature, turning on its refrigeration unit sooner rather than waiting for her temperature to climb too high. After several weeks' exposure to a hot climate, when the body has become fully acclimatized to the heat, a person can pour out two and a half times more sweat than before.

She pumps through a hamlet of four or five dilapidated houses: weather-beaten clapboard, crumbling asphalt shingles, wrecked cars propped in the yards. Families sit on sagging porches in greasy dungarees and T-shirts as she strokes past in midnight-blue-and-black Lycra shorts and skintight jersey, mounted on a streamlined titanium-alloy racing frame. They stare at her, unblinking lizards taking refuge from the midday heat beneath the shade of a rock.

She passes a sign that says 25 miles per hour. It's rusty and shot full of bullet holes. She passes an abandoned mobile home beside the road. Its door swings ajar, on broken hinges. Her temperature now begins to push toward 105 degrees, beyond the "fever of exercise" zone, into the twilight realm where concentration falters, emotions swing, hyperirritability sets in. It angers her to see the abandoned mobile home. There is such waste in this land of abundance, she thinks, unlike her country, where everything is used, and then reused, and reused again. The trailer's faded metallic skin gleams dully in the sunlight.

She keeps going. Yes, she'll take the money from this country, and then she'll leave it. Light filters through the forest around her, as if the trees are thinning. The Impressionist painter has lightened his palette into brighter, airy shades of yellow and green. She feels the terrain begin to break, to roll over, like the rolling back of a dolphin surfacing and then diving deep into the cool water. One more switchback, two more switchbacks—it can't be far. Her whole body hurts. She's hyperventilating now, great gulps of air that seem to give her no oxygen. The road bends into a gully, swings out again. She breaks out of the Impressionist-painted forest. But it's not the top— not by a mile, maybe two. It is as if a giant palette knife has scraped and gouged away the lush green pigments. Underneath is a waste-land of red dirt and brown tree stumps climbing in a vast broad slope to the barren ridge line. Waves of heat from the sun boil up-ward from the baking red dirt. The shimmering black asphalt road runs in a sinuous curve up toward the summit. Reddish water spills down a ditch beside the road like a trickle of the earth's blood. Why have they done this? How could they do this to *her*? She's come so far, and she was almost finished, and now they've put this naked patch of heat and effort and pain ahead of her. She wants to collapse off her bike, lie down in the dirt, and curl up in a ball as if she were two years old.

Concentrate! she tells herself. *Keep going! You can make it!* She imagines her wild-haired grandmother careening from room to room in their little apartment. She peers back over her shoulder down the mountain. Two turns back, she spots a flash of sunlight glinting off polished alloy. Four riders slice up the road in a tight line. She recognizes them by their jerseys: the Italian and the French and the Colombian girls, plus a German rider who's barely tagging along. Even from up here she sees how hot their pace is, how strong the pumping of their thighs—they're still sitting in their saddles— the churning, unrelenting determination of their counterattack.

Push . . . push . . . push . . . Just to the top. She can hear her coach: "Get there first and the race is yours, but if they catch you now, they'll blow right past and you'll never catch up. You'll have *nothing.*"

She leans hard into her pedals, the bike swaying under her,

as the road traverses up the barren slope of red earth. The moist heat nearly chokes her. She strains past a sign promoting the wood-products corporation that logged the mountaintop: "Putting Your Nation's Forests to Their Best Use." Dozens of bullet holes riddle its thick metal, as do the dented clusters of shotgun blasts, like punches thrown from the depths of an inexpressible rage.

She's not sure she's going to make it. She feels weak all over. Painful knots are forming in her biceps and calves and the muscles of her abdomen—heat cramps, perhaps the result of sweating out so much of the sodium from her body. Her heart pounds with great, powerful, athlete's contractions but still can't pump quite enough blood to keep all her dilated veins and arteries filled and still supply the oxygen demands of her muscles. It's as if her arteries were so many water balloons attached to a single tap that lacked the pressure to fill them all at once. Blood flow to the upper reaches of her body, and to her brain, slackens. Her vision blackens and turns fuzzy at the edges, sparking like swarms of fireflies looping through the night. A chill runs through her body; the hairs on her arms stand erect in goose bumps. Her bike begins to weave dizzily back and forth across the pavement, her body lunging awkwardly down on the pedals, her stroke square instead of rounded. Her head pounds and she feels nauseous—the symptoms of heat exhaustion.

She has no freewheel left—her gears are maxed out. She tries to weave her bike back and forth across the road to make the last steep pitch easier, as a child does when biking uphill, but she totters drunkenly and can barely keep it upright. She doesn't dare look back; she doesn't have enough strength to spare to look back. In a blur of heat and pain and sweat she crests the top, her hyperventilating breaths flinging out of her body like animal groans. The road crowns. The pain in her thighs, her arms, her back, her neck suddenly eases. Groves of shady hardwood trees cascade down the far side of the mountain, like a waterfall of shade and coolness. The road plunges through switchbacks underneath the cool flow of shade. She knows she should shift gears for the downhill, but she can't remember which way to move the lever. Her wild-haired grandmother is standing in the road beckoning her forward, pointing her downhill toward that cool waterfall of shade, toward the finish line, toward that stream of American money, just down that hill. She aims her bike straight ahead where her grandmother has pointed. Oddly, there's no one at the finish line, but she knows she has crossed it. The victory and the money are hers—yes, *hers*—but the road itself has turned, and now her bike is tumbling down an embankment, and then it is black.

It might have been different if she'd landed in the shade. Instead, the sun bores down on her exposed arms and legs and head, its heat absorbed by her midnight-blue-and-black shorts and jersey and her black sports bra as she lies sprawled beside a rhododendron bush with her bike frame on top of her.

Her heat-generating muscular contractions have ceased, but her body continues to swelter from both within and without. Hot sunlight beating on the skin of a seminude human body can add up to the same amount of heat as the exertion of walking, about 250 kilocalories per hour. Simply sitting produces about 100 kilocalories per hour. This means that if a 150-pound, seminude person were sitting in the hot sun—whether sprawled on a beach or on an Appalachian hillside—and there is no mechanism to disperse the heat either through sweating, a cooling breeze, a jump in the ocean, or whatever, that person's body temperature would rise more than one degree every ten minutes and reach heatstroke range—over 105 degrees—within the hour.

But the bike racer's furnace is burning much hotter than if she were simply sitting. Her cellular metabolism speeds up at higher temperatures, so that when her core temperature reaches 105 degrees, her cellular metabolism—her cells breaking down fuel into energy—is occurring 50 percent faster than it does at a normal temperature of 98.6. This activity adds to the heat building inside her body.

Though her skin glistens with sweat, the high humidity prevents it from evaporating and cooling her. As she lies comatose in the hot sun, her temperature soars beyond heat exhaustion, far into the realm of heatstroke. Her systems now fail so profoundly and in so many ways that the damage is difficult even to track.

105 . . . 106 degrees . . . Seizures ripple through the muscles of her arms and legs and torso.

107 . . . 108 . . . 109 degrees . . . She vomits repeatedly, and her sphincter releases.

110 . . . 111 . . . The heat begins to destroy her cells. Proteins melt and distort, and the mitochondria, a cell's powerhouse, degenerate. Her muscle tissues deteriorate. The heat sears the fine tubes inside her kidneys and kills liver cells. It swells the dendrites of her cerebellum, the part of the brain that coordinates muscular contractions, and annihilates the Purkinje cells in her cerebral cortex, the thinking part of her brain. It causes hemorrhaging throughout her body, including heart and lungs, and damages the thin sheathing inside her blood vessels. Her circulatory system responds as if her blood vessels had been cut, triggering their clotting mechanisms and causing what's known medically as a clotting cascade. Her blood begins to coagulate inside her veins and arteries.

She vomits again, the vomitus now bloody. Perforations open in her intestinal wall, and toxins emitted by spent digestive bacteria escape into her bloodstream, which may trigger septic shock in some heatstroke victims. Others die of heart failure, presumably from heat damage to the cardiac muscle.

The list of bodily destruction from heatstroke goes on—and on. Autopsies of heatstroke victims show damage to virtually every major organ. As she lies comatose and vomiting blood, her arms and legs flailing with seizures, purple hemorrhagic spots appearing on her shoulders, what is happening inside her body is not unlike the meltdown of a nuclear power plant. There is one fortunate protective mechanism her body has granted her in the face of this massive assault of heat: She's unconscious and has no idea what is happening to her. If she were now revived, she'd remember nothing.

A little more than three minutes after her collapse, the Italian and French and Colombian riders crest the pass in a disciplined line, having shaken the German. The three of them are wondering where she is, pushing hard together to catch her, although once they do, it's every rider for herself.

The derailleurs *click, click, click* in precise succession as the three nose over the top of the pass and plunge down the other side, their spokes whirring and ball bearings singing, leaning so deeply into the curves in pursuit of the Eastern European that their derailleurs almost scrape the pavement. Their passing *whoosh* stirs the bushes that line the road, the cooling breeze rushing over their sweaty bodies.

They don't notice her lying sprawled off in the bushes. The only hope for her now would be immediate and dramatic cooling. Usually this is done with ice packs, immersion in cold water, or, best of all, a cool mist of water blown over the skin. Military heatstroke victims have been treated successfully in the field with the cooling rotor wash from a helicopter that hovered over them. With prompt treatment and hospitalization, many heatstroke victims now survive, although statistics vary wildly as to just how many, and of those, how many will suffer permanent damage to the brain or other organs. One sobering study tracked fifty-eight victims of classic heatstroke who were hospitalized during the 1995 Chicago heat wave, and found that half developed kidney failure, 21 percent died in the hospital, 33 percent were "moderately to severely disabled" at the time of discharge, and a year later 28 percent more had died. Most of the survivors had not improved.

Seventeen minutes after cresting the pass, the Italian rider wins the final sprint to the finish. She flashes under a sagging banner to some desultory clapping and brakes to a stop in the village that's wedged into a tight mountain valley. She collapses off her bike under a tree. Her coaches run up to her with wet towels and thermoses of ice water. She doesn't realize at first that she's won the race, believing that the Eastern European, through some superhuman effort, has crossed the line first.

No one misses her until the awards ceremony, an hour later, after the last stragglers finish, and the racers ask each other questions: *Wasn't she in the lead? Where has she gone?* The director of the regional chamber of commerce hands the award check to the Italian. A clutch of local politicians each say a few words, but by now the locals have retired to their porches and air-conditioned trailers and the politicians don't get much exposure. The economic benefit for the region has been virtually nil except for the money spent at a few motels and restaurants back in the county seat, plus the economic benefit for the Italian racer, who plans to spend the prize money on a new sports car.

By late afternoon, the sheriff has rounded up some locals who keep coonhounds, and disperses them in pickup trucks along the road, starting at the point where she was last seen. The packs set off, sniffing up and down the pavement. Toward evening, the air temperature has dropped to a relatively cool 87 degrees. The hazy orange ball of the sun is sinking beneath the distant purple ridges when they find her lying beside the rhododendron bush. Her body is still warm to the touch of the deputy who searches for a pulse as the nuclear furnaces of the burning sun finally disappear behind the earth's shadow.

CHAPTER SEVEN

PITCHED FROM

THE VERTICAL

REALM:

FALLING

His limbs splay out to reach for holds on the rock face like a daddy longlegs: fingers of his left hand extending up and wrapping their tips in a cling grip over a nubbin of rock, left leg spread wide to edge with the instep of his climbing shoe on a projecting flake, or tiny ledge, his right foot toe-jammed for purchase into a vertical crack. All of these are secure holds. The problem is his right hand. He's trying to touch a hold shaped like a mushroom cap that hovers to the right about two feet above his head. Get past this one move—the crux of the climb—and he's got it. The top, no problem.

He has no rope. Besides his daypack, and a small bag of chalk dangling from his hip in which he dips his fingers to absorb the sweat, he wears only a yellow T-shirt and baggy black Lycra-and-cotton stretch pants, riding up high on the calf like Tom Sawyer's raggedy trousers. On his feet he wears his most essential piece of equipment: climbing shoes as delicate and closely fitted as Cinderella's slippers, thinly shod with special sticky rubber engineered to adhere to the rock.

Holding fast with his three points of contact—his two feet and his left hand—he stretches with his right hand above his head for the mushroom hold. Splayed to their maximum reach, his fingertips fall 5, maybe 6 inches short of it. He needs a bit more stretch, a little more body length. If he releases his left foot from his edging hold and brings it directly beneath his torso instead of off to the side, his body will shift into a more vertical stance. This alone might give him the few extra inches of reach with his hand.

He carefully removes his left foot from the edging hold. He swings it beneath him, letting it dangle over the void. Taking a deep breath, he extends up again with his right hand. Still short. He didn't anticipate so hard a move; he must have strayed from the regular climbers' route up the face. He regathers himself, still dangling his left foot. He's panting now. The fingertips of his left hand are tiring. His right quad, with his climbing shoe toe-jammed in the crack, bears most of his body weight. It's beginning to burn. He's going to have to do something. Soon.

He assesses his options. What he really needs to do is a dynamic move, known to climbers as a "dyno"—a move that demands full commitment. The climber releases all his holds and, with nothing at all supporting him, makes a controlled lunge like a flying squirrel for the next hold. This dyno is not far. He has to lunge only a few inches. And the handhold is big, simple to grip as he lunges at it. He knows he can make it easily. Or he could on the artificial climbing wall in the health-club gymnasium, where if he missed he would risk a drop of only a few feet to the floor. Or if he had protection—if he were tethered to another climber who was securely anchored to the rock face and could arrest a fall instantly by tightening the safety rope in a move known as a belay. But he's not at the climbing gym. There is nobody here belaying him. He looks down to the alpine meadow on the valley floor—the green tufts of shrubs, the yellow and blue splashes of wildflowers. Only two narrow ledges and 150 feet of air time between up here and down there. He's not that high, really. But he's plenty high if he misses the move.

He'll *probably* make it. He guesses he has a 60-40 chance of success. In business he has his 60-40 rule: If he estimates that he has a 60 percent or better chance of succeeding, he goes for it, and screw the consequences. He's in the takeover business, and he is hired—for a very hefty sum, of course—by investors who want to snap up huge blocks of stock and seize control of a weak and vulnerable company with the idea of restructuring it, firing employees and downsizing it, breaking it into pieces and selling it off, somehow squeezing more money out of it. In his line of work, the odds of 60-40 are daring and buccaneering—but reasonable enough. After all, it's not *his* money or his job that might be lost in the attempt. But this is different. This is his life.

He looks up again at the handhold. A dyno 150 feet off the deck with no protection. This is a loser's move. He feels it in his gut. There are two types of people in the world: the winners, like him, and the losers, a group who make up the other 95 percent of the population. Winners know when to push ahead and when to hang back, and sometimes to remain a winner you have to stay back. He tells that to his clients all the time. As for the losers, they usually get what they deserve for making stupid moves. For making moves like this one.

Back off. That's the right plan. Downclimb to a ledge or the valley floor and scout another route. Or forget the climbing this weekend. On Monday he'll return to the office and tell the guys about his close brush on the wall. They know how hard he is—physically, mentally, in every way that counts. They've seen how he has put the screws to a target company's management team when no one else had the balls to do it. They know damn well he wouldn't back off from this wall without very good reason. He's free-soloing a 5.9 climb, after all. He'll make sure they know free-soloing is the most high-stakes climbing of all.

There are no safety nets in free-soloing. In aid climbing, the most equipment-heavy type of climbing, you wedge various types of anchors into cracks in the rock or otherwise attach them to the face—items called nuts and cams, pitons and bolts—and fit them

with ropes and slings that you use as hand- and footholds, building a
kind of artificial ladder up the face. Then there's free climbing: You
and a partner climb by clinging to the natural holds on the rock and
can easily enough slip from them but have a backup safety system of
anchors and belay ropes. You climb one person at a time so you can
arrest each other in case of a fall. But in free-soloing there are no
backup safety systems—no ropes, no anchors, no belays—nothing
except you, and the face of the rock, and your own skill and judgment. It is by far the riskiest type of rock climbing, and the rarest,
and those who pursue it are considered a breed apart.

He'll make sure at the office they understand all that, and he'll
explain to them, in his knowing and expert way, the system of rating
climbs, just so they fully grasp what he was up against. A flat trail receives a rating of 1.0. A 2.0 route might be a rough trail going uphill.
On a 3.0 route, the terrain has become steep enough so you might
have to pull yourself up with your hands in a few places. When
you're on a 4.0 route, you've entered the climbing realm, and you're
using handholds and footholds, although the surface may be short of
vertical. You're into the true vertical realm—the realm of the spider,
the sheer faces of cliffs—with climbs rated 5.0 and above, sliced into
many precise gradations of difficulty from 5.1, the easiest of the truly
vertical climbs, on up to 5.8, 5.9, 5.10. All of the climbs in the 5.0 and
above range resemble scaling the sheer wall of a building; the difference lies in the texture of that building's facade—how large and far
apart the hand- and footholds are spaced, how much they demand
of one's reach and limberness and strength and innovation.

In recent decades, however, as the precise and competitive
world of serious rock climbing has blossomed into international
events, climbers' skills have outpaced the old rating system. More
numbers added to the scale now designate excruciatingly difficult
climbs where climbers lock only the very tips of their powerful fingers over the tiniest bumps of rock for handholds and "smear" their
sticky shoe soles on the slightest bulges in the wall that constitute
footholds. They often dangle like monkeys from beneath overhangs

hundreds of feet above the ground. Now there are 5.10 climbs rated into a, b, c, and d grades of difficulty, and minuses and pluses within each one of those letter grades, and beyond those are the 5.11 climbs. The system and the skill levels currently reach up to 5.14d, the most difficult rock climb that anyone in the world has "red-pointed"— completed without falling or without resting on an artificial piece of equipment. "Climbing it clean," they call it.

He's doing a 5.9 climb—not all that difficult for a decent climber under most conditions. But he's not doing it under most conditions, no, not at all, he's free-soloing it, and a 5.9 free-soloing— without a rope, without any protection, with nothing between you and the hard, hard ground except your own technique and your iron-fast confidence—that's getting to be heady stuff. That he even attempted it is surely enough to enhance rumors of his legendary *cojones*, a reputation that has served him so well in the business of intimidating the opposition. But there are other reasons he likes to climb beyond the sheer gutsiness of it. He likes the immovable feel of the rock under his hands and feet, the different textures and structures—rough rock and smooth, the ledges and flakes and elegant systems of vertical cracks. He likes solving the complex intellectual puzzle of finding a route to the top, mentally arranging the handholds and footholds into a precise, choreographed sequence of moves, and he likes the sinuous animal strength of his arms and legs as he works up the face, through his own power and grace and brains overcoming this massive obstacle thrust up by unspeakably profound forces from deep within the earth. At moments like that he is master of the rock.

Yes, he'd tell the story of his climb to the guys at the office on Monday, especially to Antoni. He'd rub it in. That shithead Antoni had promised to go climbing with him this weekend. If Antoni were belaying him on the other end of a rope, he wouldn't be here with fingertips going numb clinging to the face of the rock, worrying about making a lunge move. He'd simply lunge, and if he didn't make it and peeled off the wall, he'd yell, "Falling," and the harness would suddenly tug tight around his waist as Antoni clamped down on the rope and caught him after a fall of only a few feet. Assuming Antoni was paying attention. He'd never really trusted the guy. Then he calls up on Friday night and cancels out in order to spend the weekend with his wife and daughter. Well, fuck him. Antoni's not going anywhere. Staying home with his wife and daughter! Antoni has just joined the ranks of the losers.

To start the climb down, he probes the wall with his dangling foot. Retreat—yes, he can live with it. He's still as hard as he ever was. He can hear the thin rubber sole of his left climbing shoe scraping about on the rough, naked granite like a dog pawing at the locked back door. *Where's that left foothold?* He tilts his head down to look. *There*—slightly higher and farther away than he remembers. He reaches with his left leg, stretching his baggy black climbing pants tightly around his ropy left quadriceps. He can't quite put his left toe on the hold. In all his maneuverings, his right toe has slid a few inches down the crack where he'd jammed it. He's now slightly lower on the wall and can't reach that left foothold. He suddenly realizes that he is stuck. He can't go up. He can't go down. He's attached by only two points of contact—left hand and right foot—like a barn door swinging on two hinges from a rock face 150 vertical feet above the valley floor. "Fuck!" he whispers, stretching again for the foothold.

The first blast of panic washes through his body. His heart suddenly leaps to its maximum rate, about 160 beats per minute. His right knee, bearing most of his weight, begins to tremble, bouncing quickly up and down. "Sewing machine leg," the climbers call it, or "Elvis leg." Tense with fear, he pushed down and stood on his toe harder than he needed to, tightly contracting the muscles of his right calf. When he tries to relax them now, however, the muscles stretch. This stretching places additional tension on them, triggering the muscular reflex to contract again, then relax, then contract, over and over in a staccato cycle known as clonus. *Clickety, clickety, clickety,*

clickety goes his right knee, like an old Singer stitching its way through a long hemline or like Elvis rocking and rolling up on his toes in his black boots. His fingertips go numb. Through their battered tips, he senses them losing purchase on the tiny nubbin of rock. He'll come off the wall unless he does something fast. No rope, no protection. A second wave of panic blasts through him. His breathing, about 24 breaths per minute when he was climbing smoothly, jumps to 40 breaths per minute in quick, sharp pants of fear. He kicks and scrapes with his free foot, scrabbling for a hold on the sheer granite. Nothing. He has to do something—*fast*. He pulls hard with his left arm on his single handhold, attempting to elevate his body in order to reach the foothold.

It is in that moment, there, that his fingertips slip from the rock.

The longest fall a human has survived without a parachute is 6 miles and 551 yards—the equivalent of a three-thousand-story building—when flight attendant Vesna Vulovic tumbled to earth inside the broken-off tail section of a DC-9 that blew up on January 26, 1972, over what is now the Czech Republic. Not far behind her in epic plummets is Nicholas Stephen Alkemade, a Royal Air Force tail-gunner who bailed out, sans parachute, from the flaming inferno of a crippled British bomber over Germany on the wintry night of March 24, 1944, and fell 18,000 feet. What he recalled from his long drop through the cold, starry night sky was a sense of resting comfortably on a cushion of air and a great relief at having escaped the heat and flames. His only regret was that he couldn't say goodbye to his sweetheart, Pearl, waiting for him back in Loughborough. Alkemade's vertical idyll ended abruptly when he tore through a canopy of fir branches at a terminal velocity estimated at 122 miles per hour—the maximum velocity a falling body attains, no matter how great the height of the fall, due to the increasing wind resistance—and thudded into thick, snowy underbrush. When he regained consciousness three hours later, burned and very sore and unable to walk, he lit a cigarette from his flattened pack and eagerly began to blow his rescue whistle, more than happy to be taken a prisoner of war by a German *Kommandant* and his officers. The Germans found his parachute in the plane wreckage a few days later and verified Alkemade's incredible account, jubilantly offering him their congratulatory handshakes, slaps on the back, and good wishes for this story he could tell his grandchildren.

Overcoming the pull of gravity is our universal metaphor to describe achievement. We speak of the rise of civilizations, of royal houses, of great talents; we speak of soaring happiness. And just as universally we speak of falling to signify unhappiness and failure—failure that often results from excess pride and presumptuous meddling in the ways of the gods. Satan fell from heaven. Adam and Eve fell from God's grace. Icarus flew too close to the sun—the realm of the gods—and melted the wax on his wings. "It is the gods' custom," wrote Herodotus, "to bring low all things of surpassing greatness."

That our perception of the world is so profoundly shaped by gravity and the prospect of falling surely springs from our physiology. Humans learn at a very young age to balance precariously on two points—their feet. The danger of a fall never leaves us from the moment we tumble from the womb until the day we topple into the grave. We can fall from trees, roofs, buildings, ladders, bridges, mountains, stairs; we can slip on ice, in the bathtub, on wet floors; we can tumble from bicycles, cars, horses, carts, chariots, airplanes; we can land on earth, stone, wood, concrete, water, and—for the fortunate—a net, a mattress, straw.

Gravity follows us everywhere we travel on the face of the earth, although it's about 0.5 percent stronger at the North and South Poles than it is at the equator, where one is standing at the metaphorical rim of the merry-go-round instead of its center, and there is a slight outward-pulling centrifugal force to offset gravity's downward tug. Galileo, contrary to legend, didn't drop weights from the Leaning Tower of Pisa, but he did measure gravity's acceleration of an object falling to the earth's surface: 32 feet per second per second. This means that after each second, the falling object is traveling 32 feet per second (or some 22 miles per hour) faster than after the previous second. It didn't matter if the object was a feather or a lead ball, Galileo understood. Assuming there was no air resistance—if they were in a vacuum—they would plummet at exactly the same rate.

Isaac Newton, lingering in his mother's Lincolnshire garden after the plague had shut down his studies at Cambridge in 1665–66, speculated whether the same force that caused an apple to fall to earth extended out into space as far as the moon. Studying the orbits of the moon and planets, he arrived at the classical formula that describes the universal force of gravity: Every particle of matter in the universe exerts a force of attraction on every other; larger masses exert larger forces of attraction than smaller ones; and this force of attraction, opposite of that between lovers, weakens greatly as distance separates the objects.[1]

Galileo spent the last eight years of his life under house arrest for his meddling with gravity and the heavenly bodies, as if proving Herodotus' dictum. The brilliant Newton fared poorly, too, simmering with rage and jealousy and paranoia at his rivals' successes, punctuated by frequent nervous breakdowns. His father had died in Newton's infancy, his mother remarried a wealthy man and moved away to a nearby village, leaving him with grandparents until he was eleven while young Isaac threatened to burn down his mother and stepfather's house with them in it.

Surely Newton had a visceral understanding of the hazards as

---

1. Newton's formula is $F = G(m_1 m_2)/d^2$, where F is the force of attraction, G is the universal constant for gravitation, m_1 and m_2 are the masses, and d is the distance between them.

well as the benefits of the gravitational attraction between bodies. Without gravity, we'd all launch off the face of the earth into space; at other times—all too often—gravity hugs us far too closely for our own health. Falls are so prevalent and so universal, they rate their own listing in the World Health Organization's mortality report: about 300,000 deaths annually. Falls, like heart disease, are an affliction of the wealthy nations, killing twice as many people proportionately from rich countries (1 percent of all deaths) than poor ones (0.5 percent of deaths), presumably because the inhabitants of wealthy nations have more buildings, power lines, and bridges from which to tumble, as well as the luxury of living beyond infancy long enough to climb them.

Mountain-climbing accidents account for only a tiny percentage of falling deaths—about thirty-five fatalities annually in North America. But what they lack in numbers they make up for in metaphorical richness. Like Icarus, climbers venture close to the lofty and rarefied realm of the gods. What you make of this depends on where you stand. For those who remain below, it's easy—perhaps too easy—to interpret a climbing fall as the gods' retribution for the hubris of the climber. For the climber, the very possibility of a fall may be part of climbing's subliminal attraction. It is not an attraction to death, but the opposite—a passion for that sense of acute aliveness and total focus and full sensory alert when danger is near. Rock climbing demands one's total attention. There is no past and no future when climbing, no distant hopes for what might be or lingering regrets for what could have been. There is only the present, this moment, this movement up the rock—at times beautiful and graceful, at times awkward and difficult, but always utterly encompassing. A fall—whether a tumble quickly stopped by an alert climbing partner on the other end of the rope or an uncontrolled and fatal plummet to the earth—is the price one pays for that sense of being so acutely in the present, so acutely alive.

If for a poet a fall is metaphor and for a climber it's a kind of intensifier, from a physicist's point of view, a fall might represent the gravitational attraction of one mass for another. For a doctor a fall most likely means a broad and variable medical condition captured in two words as harsh and abrupt as the event itself: blunt trauma. Caused when some blunt object strikes the human body, or the human body strikes it, blunt-force trauma is distinct from penetrating-force trauma. The latter would include knife and gunshot wounds, while the former might occur when a tractor rolls over on its driver, or someone is thrown onto the pavement from a car that's rolled over, or, in this case, when a climber falls to the ground. In this event, the medical terminology becomes more specific—"deceleration injury." This includes the array of damage that can occur when the human body comes to a stop. A sudden stop.

There is no time to feel fear as he topples backward. Within a few hundredths of a second, his sudden release from the rock triggers his startle reflex—a response demonstrated by even the smallest infants if suddenly lowered a few inches. His arms fly outward as if to grab something, his eyeballs spin to seek the danger's source, his torso curls forward to protect his organs. The implacable tug of earth's gravity, meanwhile, accelerates his body at Galileo's 32 feet per second per second. Translated, this means he'd be falling about 22 miles per hour at the end of the first second, 44 miles per hour at the end of the next, and about 66 miles per hour at the end of the third. He wouldn't speed up as quickly, however, as he fell faster and wind resistance increased. By about the ninth second, he would have reached a terminal velocity of about 110 miles per hour in the spread-eagle position that skydivers use. In a headfirst dive, he'd reach a terminal velocity of about 190 miles per hour.

Few falling climbers hit terminal velocity, even in a long fatal fall, for they'd need to drop down a sheer face for nearly a thousand feet before reaching that speed, assuming the climber is in the spread-eagle position. What climbers have witnessed—and what a few very lucky ones have survived—is an awful kind of bouncing fall. The unroped climber slips on a steep, icy slope, or rappels off the end of a rope, or the rope's anchors pull out during what should be an arrested fall. Climbers below report how the victim's body starts to tumble, bouncing down the slope or face, perhaps coiled up in a nest of rope, bounding along like a tumbling rock in ever greater leaps, ice axes, gloves, hats spinning off, "rag-dolling"—that's the word they use—until the climber is no longer a person but an object whooshing down the mountain, and sails off some lip into the void, sometimes into an entirely different country—from Nepal into Tibet, for instance—probably dead of massive head injuries long before impact. It is, as they say, a very short story.

But this fall, climbing alone up the rock face and cursing at Antoni, is really quite prosaic. He drops 30 feet in 1.4 seconds, or the equivalent of a three-story building in the time it takes to say "How are you this morning?" He has no chance to sort out the tumbling confusion of the fall, much less watch his life flash before his eyes. There is a moment's blur of rock and sky, an instant of suspended silence. *Crack!* His right leg strikes a ledge. The blow somersaults him head over heels backward. He hears a sudden rush of wind in his ears, a rattle of falling stone, senses a churning whirl of blue sky and gray rock as though they were being beaten together in a bowl . . .

Suddenly it's very quiet and he's nowhere. He floats in the dark, suspended, indefinitely. After what seems an eternity, he hears a slowly gathering rush of sound, a clattering, like a horse and rider galloping over cobblestones toward him. He feels peppering little hits over his body, and a *plonk* onto his forehead. That was something real—falling stones. He's not dead, after all. He tries to breathe. Nothing happens, as if all the air has been squeezed from his body like a spent balloon, and it requires more effort and pressure than he possesses to reinflate it. Is it worth it? So much easier not to breathe. But if he doesn't breathe, this is the end. He will die.

He wills his chest and diaphragm again, forcing a spasm of a breath, then another and another. He becomes aware of an intense pain far in the back of his chest, between his shoulder blades, as if a hand were reaching through his chest and tearing at the flesh along his spine. Just as suddenly it subsides. Panting raggedly, he opens his eyes. He's lying on his side. He's looking out across the valley to the rock face on the opposite side and, still well below him, the alpine meadow of the valley floor.

An intense wave of nausea wells up from his gut. He leans his head over the edge, gags once, spews a slurry of maple-flavored oatmeal and $18-per-pound Kona coffee in a long arc out into the void, pants three times hard, then projectile-vomits over the edge again. After so severe a blow to the torso, his body instinctively rids itself of its burdensome abdominal contents, preparing itself for fight or flight. The human body doesn't want to be worrying about digesting breakfast when a battle is at hand.

He settles back, panting, feeling slightly better, and slowly pieces together what must have happened. After he peeled off the wall, he fell 20 or 30 feet, struck a ledge, bounced off, which helped slow the fall, fell another 20 feet or so, and landed on this ledge. It's about the size of his condo balcony, maybe 10 feet wide and 20 feet long, a few tufts of dried grass growing from the crevices at the far end. It's solid granite. He's fortunate; he landed on his backpack, which helped cushion his fall. It's a law of physics that the shorter the distance of deceleration, the greater the force of impact. With his backpack offering about 1 foot's worth of cushioning, and his body weight of 170 pounds and a fall of 20 feet, he hit the granite ledge with the equivalent of about 3,400 pounds of force. If, instead of landing on the backpack, he'd landed on, say, a thin sleeping pad that offered only one-tenth that amount of cushioning, and assuming his body didn't compress at all on impact, he'd pile onto the granite ledge after a 20-foot fall with around 34,000 pounds worth of force, or about 17 tons.

After he vomits again, gasps for breath, and regains some of his equilibrium, he slowly squirms back from the edge. Something feels odd about his right leg. He looks down. It appears that a thick, broken stick has somehow lodged itself in his climbing pants. Did he hit a tree limb during his fall? He looks more closely and discovers that it's not a chunk of wood at all but a chunk of his right femur—his thigh bone—protruding from the stretchy black fabric. He tears the fabric of his pants into a larger hole and peers inside. The end of the fractured bone sticks from an ugly red rent in his skin that exposes the shredded muscle tissue and a sticky film of blood.

Oddly, it doesn't hurt much—at least for now. What else could he have broken during the fall? A dull pain throbs from his lower left ribs. He mentally runs a check over his whole body—head, arms, back. Everything else seems okay—it's just the ribs and the leg. The pain's not even that bad. He'll be all right. He might even be back climbing in a couple of months. He guesses it's an 80-20 chance—probably even better than that—that he'll heal up like nothing happened. He definitely made the right choice up there on the crux. Yeah, he fell, but he's still alive, isn't he? That's a lot better than being dead, isn't it?

He leans his head over the edge. Pillars of granite drop to the valley floor. At least 100 vertical feet of 5.9 downclimbing, like working his way down the outside of a ten-story building. There's no fucking way. It would be hard enough with two good legs, jamming his way back down the cracks the way he came up, wedging feet and hands in crevices. If only he had a rope, he could anchor it up here, wrap it around his torso, and rappel off the ledge. He'd be standing down there among the wildflowers in ten minutes. But he hadn't bothered to put a rope in his climbing pack. What good's a rope if you don't have a climbing partner who can tie into the other end and belay you?

The cell phone! *Yes!* He'd forgotten about that! Angrily packing on Friday night after talking to Antoni, he'd tossed his spare cell phone into his climbing pack instead of his rope. Ounce per ounce, he'd figured that in the unlikely event he needed a rescue, an 8-ounce cell phone carried a lot more punch than an 8-pound rope. *Yes!* That was the kind of thinking ahead of the curve that had served him so well in the business world. He shucks off his pack, leans back against the wall with his feet toward the void, and probes into his pack, feeling for the phone's calfskin sheath. His chance of success has just jumped well up into the nineties.

Emergency-room physicians know how to calculate odds, too. They frequently refer to something called "the golden hour." Unless a trauma injury is catastrophic, for about an hour afterward the human body can more or less hold itself together and maintain blood pressure despite internal bleeding from a ruptured organ. Once the hour is past, the odds of survival start dropping precipitously. An hour, of course, is a long time if you've had a traffic accident in a major city and an ambulance whisks you to an operating room and within thirty minutes of impact you're under the surgeon's knife. An hour in the wilderness, however, is hardly the beginning of an evacuation that, especially if helicopter rescues are ruled out by bad weather or impossibly steep terrain, might take many hours or even days until the victim—if still alive—is brought to some rural outpost.

As he digs into his climbing pack, past his water bottles, his lunch, his emergency kit, his rain parka, he dimly notes the dull pain from his lower left rib cage. Blaming it on the fractured ribs, he's unaware that the impact of his fall not only broke the ninth through twelfth ribs on his left side but, just beneath them, momentarily squashed his spleen, the fist-sized organ on the left side of his abdomen that filters his blood. When the soft, blood-filled mass hit the ledge, it split open like a ripe cantaloupe dropped on the sidewalk— not a huge split, but one large enough that his spleen is now leaking his blood supply out of his circulatory system and into his abdominal cavity. There are few nerves in the spleen, however, and he feels no severe pain from the injury. He has no idea that he's slowly bleeding.

He finds the cell phone at the bottom of his pack, slides it from its elegant sheath, flips open the mouthpiece, clicks the power button, punches in 911 and hits send. He hears a click, and then a rapid *beep beep beep*. No connection with a tower; he must be too deep in the canyon. *Fuck!* He clicks the power off, clicks it on again, tries 911 again. *Beep beep beep*. He punches buttons faster: 0 for the operator. The number of his office. His old home phone number, now the number of his ex-wife. Still only the *beep beep beep* of a failed connection, as if all of cyberspace were filled to capacity with more-powerful devices and no place could be found for his lone, weak signal from the ledge.

He can still see the pimply face of the smarmy clerk at the electronics store who sold him the phone. The lying little shithead said it was the most powerful model going and guaranteed it for life. The only reason he'd bought the thing was that he'd made a pile of money buying up a piece of the company that manufactured them. He'd known that half their products were pieces of shit, but it didn't matter what *he* thought, it only mattered what the chumps out there who bought the stock on the open market thought, and you couldn't go wrong betting on the lemminglike instinct of the chumps. Sure enough, the stock began to climb. He'd figured that deal at 75-25, and he'd turned up a winner. Again.

But now he wants to fling the cell phone over the edge and watch it shatter into its shoddily made silicon components on the rocks below. No . . . he'll hang on to it and try again later. The radio signals might strengthen at night, and he can make a connection. If he does, the first number he'll call is 911 and the second will be Antoni's, to tell him what an asshole he is for doing this to him.

He looks over the edge, assessing his situation once more. No rope. No cell phone. A badly broken leg. But he'll turn up a winner again. He's sure of it. If he had to put down money, he would *defi-*

nitely, definitely place it on himself. That's what he'd honestly tell a client. No bullshit. We're talking 96, 97, maybe even 98 percent chance of success. You'd have to be an idiot to bet against him. Here's what's going to happen, he'd tell the client: On Monday morning—that's only the day after tomorrow, forty-eight hours—they'll miss him at the office. Antoni will know he went climbing this weekend and sound the alarm. A few hours later a rock climber from SAR—search and rescue—will rappel down onto the ledge from above, lower down a litter, and he'll be plucked by rescue helicopter *outta here.* That's the scenario. He's sure of it. And by Monday evening he'll be resting comfortably in a hospital bed with his leg in a cast and clean sheets and some young-babe nurse massaging oil into his chafed and bruised winner's body, admiring his finely sculpted winner's pecs.

He can make it to Monday, no problem. It's just a broken leg and a few fractured ribs. He has a liter of water, plus the cheese, apple, bread, and candy bars in his lunch. It won't be that cold tonight. Yeah, *at least* 98 percent chance of success. Probably more like 99. Antoni had better show up at the office on Monday morning . . . that's the only thing. He'd better not skip out of work like he bailed out of this climbing trip. Did Antoni say he was taking a long weekend with his family, a few extra days off? He can't remember. He was too pissed off at the time to listen. He'd jammed his thumb on the phone's off button while Antoni was still blathering away about his commitments to his family. Antoni had better be at work *early* and notice the empty office, the lack of a commanding presence in the halls, the vacuum left by his take-no-prisoners, balls-to-the-walls drive that has been a powerful engine in the firm's success. He'd *better* notice. Antoni's the key. If Antoni doesn't realize he's in trouble, then he really is in trouble. Goddamn you, Antoni, look around. Open your fucking eyes for once.

The golden hour slips away in his chattering mental monologue of calculated odds, a statistical shield of denial to keep the fear at bay. But soon he has much more to think about than odds. Once the golden hour ends, the real pain begins. His brain has been so busy dealing with the immediate aftermath of the fall—maintaining his vital functions, assessing his physical situation, bolstering his will to survive—that it's blocked out the pain signals trying to force their way up his nerves from his broken leg. It does this by releasing the body's natural painkilling substances called endorphins ("morphine within") and the extremely potent form of it known as enkephalin. Just like morphine manufactured from poppies, which it closely resembles in chemical structure, these molecules attach themselves to the nerve endings that are receiving pain signals from the broken leg. Thus they prevent the nerves from emitting substance P, the chemical that transmits the pain signal to the next set of nerves farther up his body and to the brain.

Past the golden hour, however, his brain lifts the blockade, wanting him to know exactly how badly injured he is, and telling him of the urgent need to do something about it. Shifting his back, which is propped against the rock wall, into a more comfortable position, he accidentally jars his leg. The two shattered ends of bone grate against each other. Pain impulses race through his nerves, now unplugged by endorphins, toward his brain at speeds up to 300 miles per hour—twice as fast as a jet on takeoff.

"*Shit!*" he groans.

His fingers clench into fists. He holds himself there, his whole body tensed. When the wave of pain subsides momentarily, he pulls out his emergency kit from his pack—a small plastic box containing matches, compass, bandages and gauze, and whistle—pops open the canister of ibuprofen, tosses three of them onto his tongue, and washes them down with a gulp of water. They'll block the production of prostaglandin by his damaged tissues, a chemical that helps clot the blood but also sensitizes the nerve endings to pain. But for a compound fracture of the femur, the three ibuprofen will do about as much good for the pain as a Band-Aid.

The pain rises in waves, like ocean swells before a storm. Across the valley, the golden band of afternoon sunlight slowly climbs the opposite cliff, the warm ochre hues it throws on the granite gradually replaced by gray shadow. A chill wash of air tumbles down the cliff. The rock beneath his body turns cold and hard; he looks for something softer. He has his pack, his rain jacket, and then he spots the tufts of grass sprouting from crevices on the far end of the ledge. Dragging himself slowly on his elbows, he maneuvers to the far end and lies on the vegetation. With his backpack as a pillow and rain jacket as a blanket, the thin, organic mattress provides some insulation and padding from the cold rock. But still the leg sends up throbbing waves of pain, causing him to twist facedown on the bed of grasses, digging his fingers through the bed of vegetation, scraping his nails against the rough granite until they bleed, and he crests the wave, surfing down the far side, until the swell starts to rise again.

He can't stand this. Not for forty-eight hours. He knows the angulated femur should be set—the bones aligned in their proper position—but how? He's seen old movies where the wounded hero lies in a sweaty grimace while his buddy jerks his leg straight. But how to do it alone? The thought of yanking on his leg, even if he could, makes him want to vomit. But he has to do something.

As he catches his breath during one of the troughs, he spots a rabbit skull wedged in a crevice in the rock. Perhaps an eagle deposited it there, or maybe it somehow tumbled from the top. The skull stares at him with empty eye sockets.

"Stop looking at me," he hisses at it, panting.

It stares back vacantly.

He stretches his arm over to the crevice, grabs the skull, and flings it over the ledge.

"Get outta here, you nasty little fucker."

The effort costs him another nauseating wave of pain but reveals his opportunity. He slowly wheels himself on his back, like a turtle spinning, until his feet lie near the rabbit-skull crevice. He tears the thin fabric of his climbing pants open further, assessing the angle of the bone jutting out. The lower shaft of the femur overlays the upper shaft by about two inches. As he watches, his powerful thigh muscles involuntarily spasm into contractions, and he can see the shattered bone ends shift and subtly grate against each other, triggering another overwhelming wave of pain.

Clenching his teeth, he hoists the leg and wedges the gummy sole of its climbing shoe into the crevice. Bracing his left foot against the rock, he lies back on his elbows, takes three deep breaths, and, as if he were stomping his foot in anger, shoves hard off the rock with his good leg.

"*Uuuuuunnnngggghhh,*" he exhales, a guttural string of consonants of effort and pain. He can feel the thigh muscles stretching.

There's a grating and a click. The projecting end of bone suddenly disappears back into the ragged, red aperture of the wound. The pain eases, as if the pounding of the surf suddenly flattened into calm gentle seas. He lies back, exhausted from the effort, his foot still wedged in the rock. It feels so much better here. He decides to leave it wedged, pulls over a small boulder that's lying on the ledge, and props it carefully under the back of his knee to hold his leg in the raised, stretched position. Without really knowing it, he has put his leg in traction—the correct treatment for a broken femur. The steady pull of traction stretches the spasmodic, contracted thigh muscles back to normal length and closes the veins and arteries, stopping further bleeding. Relieved of the torment, and not eager to jar the leg, he searches in his pack for something with which to splint it. Finally, with his pocket knife he slices a piece from the foam backboard of the pack itself, carefully wraps it around his leg and fixes it with the pack's cannibalized straps.

He lies back, panting slightly, pleased with himself. Making it to Monday's no problem now. He's back up in the 97th, 98th, even 99th percentile. Just stay calm, ration the food and water, get some sleep, and before he knows it some young search-and-rescue guy—who knows, maybe it will be a babe—will be rappeling down and landing on his ledge with a cheery "How ya doin'?"

He'll act nonchalant, like it was nothing to be out here two nights with a broken leg. "Got stiffed by my climbing partner," he'll say. "So I'm just sitting here enjoying the view."

That ought to impress them. Yeah, things are looking good. The leg feels so much better. Now it's only the ribs and the odd pain—a kind of pressure from inside, really—just beneath them. He runs his fingers over the tender area. Probably bruises, he tells himself. Still okay. Still up there in the high nineties. Still a winner.

He lies back again on his grass bed, very tired, still panting slightly, hearing the pounding of his heart in his ears. It's now well into the third hour since his fall. The blood slowly leaking from his spleen has formed a pool so large—as large as a grapefruit—that it presses out against his peritoneum, the membrane sack that surrounds his internal organs, causing him to feel the odd pressure. He's lost about two quarts of blood out of his total of about five quarts. His heart must pound faster to speed the diminished supply through veins and arteries in an attempt to maintain blood pressure. His heart rate jumps, more than doubling from his resting runner's pulse of 55 beats per minute to 130 beats, but his heart is losing the battle and can't keep the blood vessels adequately supplied. His blood pressure sinks from its normal of 120/80 to 80/50—the first number in the pairing, his systolic blood pressure, referring to the pressure of the blood his heart squeezes out into the arteries during its contraction, while the second number, the diastolic pressure, measures the arterial pressure during the heart's relaxation phase. Like barometric pressure, both are calibrated in millimeters of mercury that this amount of pressure would force up a glass tube containing a vacuum.

He slides into a state of shock. Not the layman's version—that emotional daze in the immediate aftermath of a trauma—but the medical version, hypovolemic shock, or low-blood-pressure shock. In someone who has a heart condition or is elderly, this drastic fall in blood pressure could trigger a heart attack or stroke, but he is relatively young and healthy and the drop alone is not life-threatening. Without adequate blood pressure, however, his muscles and organs don't receive the oxygen they need. He begins to hyperventilate like the wounded cowboy hero in the Hollywood shootout. His oxygen-starved muscles ache all over his body. He craves water as his thirst sensors tell him to replace the fluid lost through internal bleeding. From lack of oxygen, his thinking blurs into a timeless drift, floating on the slowly fading ochre light, hovering beneath the stars that appear in the purple sky. A sense of anxiety and doom—a physiological response to very low blood pressure—comes over him.

He might die here. The thought is an abstraction to him. He tries to calculate the odds but comes up with double zeros, the numbers pushed aside by the simple thought that he might die here, alone on a ledge suspended above a green alpine meadow. What does it mean? He doesn't know. Why in this way, on this ledge, of all ways? It doesn't really matter. Has he lived a good life? He clumsily tries to wrap his mind around the sum total of his actions, but he can't seem to manage it. He still possesses the bluntness to ask the question but not the mental agility to answer it. He drifts in a vacuum, as if his supply of ready justifications has deserted him, suddenly leaving him without anchor points to the world that he knew, without a rope, without a belay, hovering above the surface of his own life, his head spinning with vertigo. He can't think what he would or would not do differently. He can't think at all.

By the fifth hour, his ruptured spleen stops bleeding. Enough blood has leaked out into the surrounding tissues that the growing lump of blood-infused tissue, known as a hematoma, presses against the puncture and essentially plugs the leak. It's helped in this by his low blood pressure—the body's last line of defense against a major wound—as if it slows all systems in order to repair itself or to simply buy time. His condition begins to stabilize, although at an impaired level. His heart and breathing rates remain high, but his consciousness slowly clears, especially when he lies quietly, burning little oxygen. He sits up briefly, dizzy and light-headed, craving water. He forces himself to eat one of his candy bars, masticating as if it were a mouthful of dry leaves, carefully drinks a few mouthfuls from his precious liter of water, then lies back again beneath the black sky.

He wakes only once during the night, hyperventilating slightly, heart still pounding, and lies quietly on his back, his head somewhat clearer, and stares up at the icy bright stars. If he dies on this ledge, only the stars will watch him vanish. Who will miss him? His parents are dead. His wife has divorced him. He has no children. Would there be a funeral? They'd do something for him at work. All the money he made for the firm. Or would they just shuck his memory aside like it was another asset they'd acquired and no longer needed?

The warm tears gather at the corner of his eyes and trickle down his temples. The stars of Orion, sword dangling from his belt, spin slowly overhead, until he finally lapses again into sleep.

He wakes at dawn, under a pinkish sky. Slowly, methodically, he pulls himself to his elbows and checks his condition. He feels as though he could drink a gallon of water in one, long gulp, but other than that, he seems better. He moves his good leg. He examines his bad one. No blood from the break. He feels his pulse—it's slower, and he's breathing easier, as if the quiet rest of the night helped restore oxygen to his system. The odd pressure still pushes from beneath his ribs, but he seems otherwise stable.

He feels pretty good, all things considered. Back up there in the 95th percentile, even higher. One more night and he's home free. The night wasn't bad at all, except when he woke up, thirsty again, and started thinking too much. *Shit!* He's getting soft on himself. Survival of the fittest—that's the way the world works, the way it always did, and he's one of the very fittest. He's made sure of it. Damned sure of it.

The sun now cracks over the top of the opposite cliff, a brilliant white disk shooting the warm rays of life down onto his ledge. Yes, he made it. We're talking 99.9 percent certainty. At least. Yes, still a winner.

"I'm still here, you sons of bitches!" he screams out into the void. "I'm still here and I'm alive!"

It would have been far better at that moment, for his sake, if he'd simply kept quiet. Fifteen hours earlier, when he landed on his side on the granite ledge, his torso suddenly stopped falling on impact, but his heart continued its momentum for a short distance inside his chest. The heart moved freely except for one part—the aortic arch, the thick blood vessel running upward out of the heart to supply the upper body. It is anchored both to the heart and to the spine. If the heart moves inside the chest too far and too fast, the aortic arch can tear or sever where it connects to the spine—one of the deceleration injuries that strikes drivers in severe auto accidents who are thrown very hard into the steering wheel, dying almost instantly.

This was the severe pain he felt far in the back of his chest on impact. The trauma of his fall didn't sever the arch's walls entirely, however, but split the middle of the wall's three layers, like the delamination in the sandwiched layers of a piece of plywood or cardboard. The fall "dissected the media" of the aortic arch, in medical terminology. For the last fifteen hours, blood continued to flow through the aortic arch in a "false channel" between the layers of the wall, all the while forming a growing bulge called an aneurysm. As the aneurysm grew, which they tend to do over time, the walls weakened. It was good for his aneurysm that he'd had a quiet night. It was very bad for his aneurysm that he now started screaming victoriously, angrily, defiantly, into the void.

Adrenaline pumps through his circulatory system. His heart rate, which had been down around 90 beats per minute, climbs to 120. Worse, his blood pressure jumps from 100/60 to 140/100.

"Wake up, Antoni, goddamn you!" he screams, the words echoing between the cliff faces. "Here I am!"

When he screams, he superinflates his lungs, bringing up his diaphragm to generate the air pressure for his shout, which increases pressure on the large blood vessels in his chest, which backs up the blood in his veins. This causes the jugular vein to stand out boldly in his neck in the manner of angry, red-faced bosses in the midst of a screaming fit. It also, unfortunately for him, forces his blood pressure higher still.

It's then that the aneurysm ruptures. He feels an odd blow deep between his shoulder blades. The aortic arch splits open. A pulsating river of blood gushes with every powerful pump of his heart not to his upper body and brain but directly into his chest cavity.

He doesn't even have time to get off one more shout at the sons of bitches out there. Within ten seconds he loses consciousness. He crumples backward onto his grass bedding. His heart gives a last hundred beats, emptying his blood supply into his chest cavity. Then it stops.

It was just as well for his sake that it happened as it did. Antoni didn't come back to the office that Monday morning, away with his family for a few days' vacation. When the search did finally start for him, the firm called in his sister from a thousand miles away, and Antoni told the police and the search-and-rescue team where they'd planned to climb. The searchers quickly found his new SUV parked at the trailhead. But then the trail went cold. He wasn't on any of the climbing routes that he and Antoni had discussed, nor on the ones nearby, and he hadn't told anyone else just where he was headed. People speculated whether he was a victim of murder, or was a suicide, or had perpetrated an embezzlement scheme. The firm checked and double-checked its accounts, but found no funds missing.

He was finally discovered by a hunter who was glassing the cliffs from above for mountain goats and saw something strange. He's found in the exact position in which he collapsed with his last shout, leg still wedged in the rocks, but it's six years later now, and of course all that's left is a skeleton picked over by the birds but for a few unappetizing tatters of Lycra and cotton, plus a cell phone, its black plastic weathered gray by sun and rain and snow, lying on the ledge beside the bones of the outstretched hand.

CHAPTER EIGHT

THE SHARP STING

OF PARADISE:

PREDATORS

"Smell this!" Mary said, pressing the flower toward Gil's nose. "I wish we'd had some of these at the wedding."

He sniffed cursorily and turned his face away. "Let's go find a beach."

"Why don't we stay here awhile first?" Mary said. "We can find beaches anytime. But this"—she waved the flower up toward the leafy, cathedral-like ceiling of the rain forest's canopy, the thick vines dangling from the arching branches, the orchids sprouting from tree trunks, the tangled profusion of life—"this place is *extraordinary*! This looks like the land before time!"

"I'm heading back to the car," Gil said, shifting his sandaled feet over the forest floor as the ants crawled up between his toes.

"How can you expect to appreciate the beauty of a place if you refuse to spend some time in it?"

"I don't consider this place beautiful," Gil said. "I think it's malevolent. And besides, there are ants. Let's go."

She dropped the hand holding the flower down to the hip of

her khaki shorts. Her hand just missed brushing the leaves of gimpi-gimpi or stinger plant, whose hollow hairs inject poison into the skin that can itch and burn for up to six months after contact. Meanwhile, a taipan, one of the world's deadliest snakes, had curled up to sleep a few hundred steps away.

"You always want to be somewhere else, Gil," Mary said. "Why can't you just enjoy where you are?"

"As if you're never in a hurry," Gil said. "I seem to remember you were in quite a rush to get married."

"And you certainly didn't waste any time walking out on your ex."

But he'd already started to walk back along the overgrown path toward the car, angrily shaking his feet as he went to dislodge the ants.

They drove in silence north up the coastal highway. Dark clouds had bunched over the coastal range on their left and rain began to spit against the windshield of the rental Jeep. To the right, through breaks in the low forest that fringed the beach, they caught glimpses of the sea—calm and blackish green as the rain peppered the surface. The rainy season—"the wet," as people called it here in northern Australia—was almost over. They knew the shower would last only a few minutes before the sun broke through again. They'd planned their honeymoon to take advantage of exactly this month, April, on the cusp between the wet and the dry, having flown halfway around the world to spend two weeks amid the exotic flora and fauna of Queensland's Cape York Peninsula and the area nearby. The peninsula was a lush and beautiful spit of land that projected from the north coast of Australia into the tropical waters of the Coral Sea like a sharp spine about to puncture the underbelly of the great bird-shaped island of New Guinea.

Gil held the wheel steady as the car planed through a large puddle, the water thunking up against the floor boards. They didn't see the saltwater crocodile lounging in a wet spot—a billabong—near the road.

"When do you want to go out to the reef?" Gil asked.

It was his way of offering to make up.

"How about tomorrow?" Mary replied. "We could go to the beach today and head out to the reef tomorrow. We can ask the hotel desk clerk tonight to arrange a boat."

It was her way of accepting his offer. They both knew that later in the day, back in their room overlooking the sea at the big resort, their lovemaking would be particularly ardent.

Still, the honeymoon had been rough. They seemed to fight at least once every day, followed by long, dripping silences. It hadn't been like this back in the States. They'd met at a bird-watching camp on the New England coast when her first marriage had just ended and his was about to end, a victim of too much time devoted to his law practice. They'd each, in their own way, repaired to the seashore to sit on the dunes and watch the birds wheel about on the sea breezes as a means to restore some balance and harmony to their own lives. By the camp's end, they'd arranged their first "date": a four-day snorkeling expedition to the Bahamas. It only seemed appropriate, when he proposed to her at sunset on the beach at the end of the fourth Bahamian day, caught up in the headiness of it all, that for their honeymoon they should choose one of the world's most beautiful wild places: the Cape York Peninsula.

"How about here?" Gil said, slowing the car where a sandy track led from the highway toward the beach.

"It looks fine," said Mary.

Gil swung the Jeep onto the rutted track and they bounced along for a few hundred yards until it ended in deeper sand and an open forest of palms and eucalyptus. They climbed out of the Jeep, picking up their daypacks laden with cameras, sunscreen, water bottles, and towels, and walked to the beach, watching where they placed their feet. They'd been warned that, basking in the sun or hidden in the undergrowth, the death adders looked like driftwood.

"Oh, Gil, it's spectacular!" Mary said as they stepped out on the beach.

The sun was just breaking through the wet clouds that clung to the headland—richly textured with its green mat of tropical foliage—at the beach's far end. The curving strip of yellow sand glistened, and the water radiated a vivid aquamarine. They couldn't see another sign of human life along the 3-mile crescent. Gil wrapped his arms around Mary and hugged her.

"I'm so happy to be here with you," she whispered into the soft blue fabric of his T-shirt. She was quite sure she meant it.

"And I with you," he said into her sweet-smelling, honey-blonde hair. They'd get used to each other eventually, he figured. It would probably work out all right.

They dropped their packs, pulled out their towels, and spread them on the warm sand. They unfastened their shorts, slipping them down tanned legs. Underneath he wore a fashionably baggy, fashionably faded blue bathing suit, and she a small black bikini bottom. She was proud of her body and, as she moved into early middle age, worked diligently to keep it in shape.

"The water looks lovely," she said. "Let's go for a swim."

"I don't know about swimming here," Gil replied, sitting down abruptly on his towel. "There aren't any enclosures."

Back at the popular beaches near the resort town he'd seen people swimming inside what the locals called "stinger nets." These were floating, corral-like enclosures made of fine netting designed to keep out jellyfish. As he'd strolled along the beach, he'd read the sign posted at one enclosure: "Warning: Marine stingers are dangerous October to May." The sign had then listed the emergency treatment procedures for a severe box jellyfish sting.

"I saw people swimming outside the enclosures," Mary said. "They say the worst of the jellyfish season is over."

"I still don't think it's a good idea," Gil said.

"Gil, how can you let this incredible blue water go to waste? Here we are in paradise, and you're sounding like a lawyer."

"I'm just trying to be prudent in a place where we don't know the score."

She stripped off her white T-shirt and let it drop to the sand.

Underneath she wore a strapless black bikini top. She started walking toward the water, as much propelled by her defiance of his oppressive caution—of his whole being—than drawn by the tempting blue sea.

"You're being foolish, Mary."

"Whatever it is I'm being," she called back over her bare shoulder, "it's better than what you are."

H_{omo} *sapiens* is, by far, the most lethal predatory species on the face of the earth. Instead of sharp teeth, lacerating claws, or death-inducing venom, the species is equipped with powerful brains and nimble fingers with which they have learned to fashion clubs, spears, knives, axes, pit traps, blow darts, M-16 rifles, land mines, napalm, nerve gas, supersonic fighter jets, cruise missiles, and neutron bombs. *Homo sapiens* preys on and kills over 1 million humans annually—about 940,000 through war and another 200,000 through violent crime. The next most lethal animal, snakes, by comparison kills only 65,000 humans per year, or less than 6 percent of the number *Homo sapiens* kills. The third most lethal animal, the crocodile, dispatches only about 960 humans annually, and the fourth, the tiger, claims roughly 740 human victims each year. The much-feared shark falls far down the list—only about 9 human victims annually worldwide—making it a lightweight compared to the ostrich, which can kick viciously with hammerlike feet and sharp talons when cornered and kills about 14 people every year. As for the ferocious grizzly bear, it ranks slightly below the puny, ratlike mustelids (weasels, badgers, skunks, ferrets, etc.), which kill about 4 humans a year, primarily pet ferrets attacking unattended human babies.[1]

---

1. "Human Death from Animal Attack 1978–1995," data gathered by Remote Care Management. S. A., Baily, S. A., Ishiakara, M. V. Callahan. War and homicide statistics vary depending on year and source. World Health Organization

Who would guess that in the United States and Canada you have more to fear from the moose—that seemingly benign creature immortalized in affable cartoon characters—than any other creature, with about six human deaths annually, compared to about five for snakes? When Bullwinkle, who can weigh nearly a ton, has sex on his mind, he likes to mix it up with his love rivals, for whom he might mistake you, using his antlers for weapons. Even gentle Bambi claims about one human each year in the United States and Canada. And here are a few more reassuring statistics: The most likely place in the United States to be attacked by an alligator (three deaths between 1992 and 1998) is not deep in some swamp but on a golf course. Your chances of being attacked by a shark while swimming along the coast of North America are approximately one in five million, although there is an area off the coast of northern California known as the "red triangle" where great white sharks occasionally mistake wet-suit-clad surfers who are paddling their boards for basking elephant seals, which happen to be one of their favorite foods.

Very few animals stalk humans as prey, and those that do, such as the infamous man-eating tigers of India or lions of Africa, tend to be individual animals that for some reason lose a fear of humans and develop a taste for their flesh. Animals, for the most part, attack humans when they are surprised by them and feel threatened, when their territory is invaded, or when defending their offspring. Snakes kill far more humans worldwide than any other animal, but, as one authority emphatically states, snakes *"have never been shown to attack without provocation despite lengthy historic commentary to the contrary."* Most bites occur when rural villagers unwittingly disturb a snake, often by stepping on it in the darkness; in the United States, by way of contrast, many victims know the snake is venomous, are under the

---

statistics for 1998 show 736,000 deaths from homicide and violence, and 588,000 from war; for 1999 they show 527,000 deaths from homicide and violence, and 269,000 from war. *World Health Report 1999* and *World Health Report 2000*. Geneva, Switzerland: World Health Organization, 1999, 2000.

influence of alcohol or drugs, and, in the words of one expert, are "messing with it." Researchers in Alabama have noted a statistical drop in venomous snakebites among adult males when University of Alabama or Auburn University football games are televised locally, presumably because instead of being out somewhere looking for wild snakes or messing with pet ones, the would-be victims are ensconced safely on the couch with a cold beer in hand.[2]

Humans have long been fascinated with creatures more powerful than they, although for equally long they have remained undecided whether to worship the beasts or destroy them, as if the very idea of a being on this earth more powerful than them offends the reach of ego and lies beyond comprehension. The Egyptians worshiped a crocodile god, Sebek; the native peoples of the northern forests honored the bear, offering elaborate apologies to its relatives before killing it; Amazonian tribes elevate the jaguar as a mythical beast. In mythology, powerful beasts can create the world or they can destroy it. Like the gods themselves, the beasts possess the sacred power to breathe life into humans and to take it away. They can—in the form of monsters such as Beowulf's Grendel, emerging from the swamp in the night to attack King Hrothgar's great mead hall—represent chaos and darkness lying just beyond the edge of the known world. They can embody sin, such as the serpent in the Garden of Eden. They can pose a crucial barrier one must overcome to earn one's manhood, such as the slaying of a Masai's first lion or an Inuit's first polar bear. What they almost never represent in the human imagination is what they almost always are: creatures not looking for trouble but, like everyone else, simply trying to survive in the world.

There are no statistics showing that one region of the world is

---

2. Unpublished observations, M .V. Callahan and R. M. Pitts cited in "Field Recognition and Management of Exotic Snake Envenomation," lecture by Michael V. Callahan, M.D., Wilderness Medicine Conference, Keystone, Colorado, July 2000.

more dangerous than another for animal attacks. Still, one can speculate. It would seem that parts of Africa inhabited by big-game animals—tiger, lion, elephant (an unpredictable herbivore that's been known to stalk humans), hippo, and Cape buffalo—would make the list, as would the snake-infested Amazon basin and parts of Southeast Asia such as Vietnam's Mekong Delta, where, in a rice-farming region hardly larger than Rhode Island, researchers have found that cobras, kraits, and vipers kill over 2,700 adults and children each year.

On the list, too, one expects, would be the northern coast of Australia and the region around the Cape York Peninsula. Besides its assortment of venomous snakes such as the death adder and tai-pan, its viciously stinging plants, and its poisonous fish and sharks, it is also home to the estuarine or saltwater crocodile, *Crocodilus porosus*—locally known as a "salty" to distinguish it from the more innocuous "freshie." Salties and their cousins throughout tropical Asia account for more human deaths annually than the African crocodiles who wrestled their way to fame in Tarzan movies. Salties can grow to be over 20 feet long and can keep pace with a speedboat in the water and with a horse on land. Their jaws are powerful enough to snap the propeller off an outboard motor. Hiding at water's edge, a salty knocks its victim—kangaroo, cow, human—into the water with its powerful tail, seizes a limb, and performs a "kill spin," revolving over and over underwater and mangling the limb until the victim dies of bleeding or simple drowning. One saltwater croc, according to re-ports, when cut open was found with the remains of an Aborigine in its stomach as well as a four-gallon canister containing two blankets.

But of all the crocs, snakes, and other faunal hazards of north-ern Australia, one creature, despite its diminutive size, towers above the rest. This is a small, graceful-looking jellyfish not much larger than a grapefruit, known by the scientific name of *Chironex fleckeri*—sometimes generically called a box jellyfish or sea wasp. Venomolo-gists regard *C. fleckeri* as perhaps the venomous creature most lethal to humans on the face of the earth. The venom itself—still largely a mystery to researchers—may prove to be among the most toxic substances known. At least sixty-three people have died along the coast of northern Australia from *C. fleckeri* or similar stings in cases documented since about 1900. *C. fleckeri* or related species also probably cause many unrecorded deaths in Southeast Asia and especially in the Philippines. Compared to the Atlantic Ocean's Portuguese man-of-war—not a true jellyfish—whose sting is painful but rarely fatal, *C. fleckeri*'s sting is in a class of its own. By some estimates, it can kill a human in less than a minute.

Years ago, no one knew what invisible presence along these coasts caused its victims to suddenly scream and thrash in the water with the instant and excruciating pain of a sting and sometimes die. The species itself wasn't identified until a boy's death by jellyfish sting in 1955. A Queensland radiologist and amateur naturalist, Dr. Hugo Flecker, urged police to net the beach in order to find the offending creature that eventually was named after him. For a time it was known as a sea wasp, but tourists from other parts of Australia began to show up on northern Queensland's beaches wearing hats and umbrellas to repel what they thought were flying insects, and so the name was discouraged. Now, during the wet season, when *C. fleckeri* congregates near shore, swimmers at popular beaches stay inside netted enclosures designed to keep it out. Researchers have developed Lycra suits to guard swimmers against the stings and a box jellyfish antivenin by injecting sheep with nonfatal doses of the creature's venom and extracting their blood products. Quickly administered, the antivenin does much to ameliorate the jellyfish's sting. Resuscitation using CPR has also proved effective in reviving victims. *C. fleckeri* stings are now relatively rare compared to stings of other, less potent jellyfish, though they still occur. The effectiveness of all the measures and education put into place to prevent box jellyfish stings depends, of course, on the measures being followed and obeyed.

"Come on in," Mary shouted, up to her knees in the clear water, waving to Gil. "See, there's nothing to be scared of."

"I'm not scared," he called out to her. "I'm prudent."

He was lying back, propped on his elbows on the beach towel that he'd spread on the warm sand, and watching her through his tortoiseshell sunglasses.

"You're going to let your new bride swim all alone?" Mary said.

"I'm admiring my new bride from afar," he called back.

Mary turned and waded deeper. The rippled sand deliciously massaged the bottoms of her feet, and the warm tropical water soothed her skin like a mineral bath. She now stood as deep as the middle of her thighs. Calm and blue-green, the sea extended like a huge placid lagoon toward the Great Barrier Reef, some 30 miles offshore. She badly wanted to see its beautiful coral and spectacular fish. It was hard enough to convince Gil simply to go for a swim; how difficult would it be to drag him out there? He'd become so unadventurous since the honeymoon began, as if this profusion of life—the tropical rain forest with its spectacular butterflies, birds, and wildflowers, and the coral reefs, with their rich sea life—that filled her with elation unnerved him, and he took refuge from it in their air-conditioned hotel room with its satellite television and minibar and on the local eighteen-hole golf course. She waded a little deeper. The water nearly ringed her hips. She knew she wasn't being prudent, but the water looked fine, and his annoying, persistent caution made her want to wade deeper, away from him. Would this be their life together, she wondered—she ready to explore, meet new people, while he dragged behind, unwilling to let go of the safe and the familiar? Was this a mistake, this marriage? Already the fights she had with Gil resembled the same shopworn arguments she'd had with Tom, her ex-husband. He would have loved the resort, too, and found few reasons to venture beyond its manicured grounds that kept the forest and its creatures at bay. A real estate developer, he would have seen beauty—and a lot of risk—in the boldness it took to carve a resort from this remote reach of Australia. To her, no matter how pleasant its amenities, it was a desecration of the primordial landscape. If Gil turned out like Tom, cautious and unimaginative, interested mainly in money, she'd leave him. Maybe first have the child she wanted, and then leave him.

"Is that all I can expect from my new husband, admiration from *afar*?" she called out, twisting halfway around toward him. "The water's delicious."

He waved her off. "I'm perfectly happy sitting right here," Gil said.

But he wasn't perfectly happy. How could he be happy with someone who was pushing him all the time? A constant, tiny *shove shove shove*. This was supposed to be vacation, wasn't it? He pushed himself hard enough at work. Every day he went out and laid his reputation on the line. It wasn't glamorous, doing corporate tax work, but it was demanding and extremely exacting, and you couldn't afford mistakes. The government regulators could smell the faintest trace of blood in the water—the tiniest vulnerability that might be hiding much deeper problems—and they'd come charging in for a feeding frenzy. Yes, it was a hostile world out there, and he didn't need someone pushing and telling him to loosen up. Mary now reminded him of his ex-wife, Betsy, who for five years had chided him to work less, travel more, go for hikes and picnics and visits to galleries, invite near-strangers to their house for dinner. He told her he wasn't *opposed* to these things; he just had to concentrate on the task at hand, namely, his career. She wouldn't let up. He'd finally moved out. Then, strangely, he found himself taking up some of the activities he'd resisted for so long, as if to prove to her that he wasn't the stodgy person she thought. It had been a great relief to meet Mary at the bird-watching camp—she was so patient and understanding when he told her about his failed marriage—but now it seemed he was reliving those five bad years all over again. Couldn't she just give it a rest? Relax a bit instead of charging from rain forest to reef to outback? He didn't want to think about how it would be when they got back home. He'd already made the mistake. A stupid mistake—the kind he always chided himself to avoid in his corporate tax work. How long would a divorce take? And, more important, what would it cost him?

He looked out over the blue-green water where she stood hip deep, her black bikini bottom low on her hips, the curve of her back sinuous, a crescent of the roundness of each breast just visible along her sides when she raised her arms. He had to admit, she had a great body.

"Last chance!" she called, twisting her head halfway around toward him.

"Hurry up and swim if you're going to swim," he called back, his irritation rising. "Otherwise let's go back to the hotel."

She brought her arms over her head and sprang gently off the sand with her toes. As she dove underwater, sliding in with a gentle splash, she made her decision: *That's it. We're finished.*

About a year earlier, late in the wet season, two spawning *Chironex fleckeri* released sperm and egg into a river estuary not far from Gil and Mary's isolated beach. Joining in the warm water, they soon grew into a minute ball of cells—the planula, as this stage is known—and dropped to the river bottom, attaching to the underside of a rock. There, through the dry season, the planula sprouted the beginnings of a crown of tentacles and grew into a tiny polyp—a kind of protojellyfish—that by the end of the dry season had metamorphosed into a small medusa, or what's commonly known as a jellyfish. Just ahead of the monsoon rains, the *C. fleckeri* propelled itself out of the estuary and into the Coral Sea.

For the next few months it gently pulsed through calm waters along the coast, avoiding violent currents and waves that could tear its delicate tissues, feeding and growing. Its body, an almost transparent milky white that consisted of 95 percent water, developed into a graceful bell shape with a squarish, four-cornered bottom rim—thus the name "box jellyfish" to designate *C. fleckeri* and about fifty related species in the order Cubomedusa. By filling its bell with water and squeezing it out like an umbrella opening and closing, the jelly could jet along at speeds up to 4 miles per hour. A chicken-foot-shaped projection called a pedalium sprouted from each of the four corners of the bell's rim, and from these grew the jelly's tentacles—up to fifteen tentacles from each foot. Only a 1/4 inch in diameter, and on a large adult jelly stretching over 10 feet long when fully extended, the tentacles formed ringlike segments and resembled twisted lengths of skinny—and highly charged—electrical conduit.

Using primitive eyes to help it avoid bumping into large objects, the jelly followed schools of shrimp that congregated just off the sandy beaches. Sensitive to strong sunlight, the jelly lingered near the shallow bottoms of the sea during the height of the day, rising toward the surface as the sunlight weakened in late afternoon, dangling its long tentacles behind it, trolling for its prey, often moving into the shallower water near the beach on an incoming tide. As it waited, a shrimp came past, extending and fanning its tail, swimming backward. Inadvertently brushing one of the jelly's tentacles, it died almost instantly. Like some mariner cowboy roping a calf, the *C. fleckeri* reeled in the shrimp with its tentacles, finally feeding the meaty morsel into wide, grasping lips.

The sea was calm, the tide rising, the sun settling down into the latter part of the afternoon as Mary put her hands over her head to make her dive. Virtually invisible to her, the jelly was swimming along 12 feet seaward of her in water 5 feet deep, its tentacles trailing behind for shrimp.

Mary never saw it.

Plunging beneath the smooth surface, she glided underwater, savoring the trickling sound in her ears and the smooth, wet warmth of the tropical sea flowing over her bare skin, extending her glide, kicking along underwater, enjoying it, relieved that she'd made the decision to leave Gil. Then something brushed against her arms. She flinched, startled. Was it seaweed? Then it brushed against her shoulders, her midriff, her back.

Reflexively, the jelly's tentacles contracted, piling themselves in loops and S-shapes onto her skin in its instinctive attempt to apply maximum tentacle surface area—and thus maximum venom—to its victim. Each tentacle was embedded with millions of tiny sacs of venom called nematocysts, the entire jellyfish armed with an estimated four billion of them. The touch of Mary's skin—indistinguishable, as far as the jellyfish was concerned, from a shrimp or a fish—bent down a triggerlike spike known as a cnidocil that projected from the top of each sac, springing open the nematocyst's lid like a jack-in-the-box. Instead of a clown, however, inside was the equivalent of a hypodermic needle—a tiny coiled tube with a sharp pointed tip. The tubes now launched out—thousands of them—extending about .03 of an inch, and jabbed their tips into Mary's skin, injecting venom from the sacs into the tissue and capillaries of the dermis just beneath the outer skin surface. Meanwhile, a wrap of tiny spikes around each tube's base snagged her skin like a coil of barbed wire to fasten the tentacles securely to her body.

Mary gasped underwater. A stream of bubbles escaped from her mouth. The stinging swelled into an incredible mounting surge of pain, burning over her bare arms, her shoulders, back, torso. No, she thought, not seaweed. A thousand hornets had embedded their stingers in long, pulsating strips across her skin.

She kicked and arched toward the surface, tearing at the stinging things adhering to her skin. She held in a scream of pain, the bubbles spurting from her nose and mouth. *Get your head up first,* she thought. *Get your head up.*

Gil, leaning back on his elbows and staring with irritation through his tortoiseshells, watched her graceful dive and then the long, suspended moment of her underwater glide. The surface erupted in a bubble of thrashing foam. He sat up. It was Mary's head, her sun-streaked hair wet and sleek, her arms flailing wildly at the water around her, clawing at her chest and back and shoulders at something that he couldn't see. Her screams and grunts of pain carried across the calm surface, short and gutteral at first as she plucked and tore at her skin, then sustained and high-pitched as she tried to twist toward shore for his help.

He knew immediately what had happened. *Why hadn't she listened?*

"I'm coming!" he shouted, launching himself off his beach towel and onto his feet

He would be the heroic husband, vindicating himself for the caution that she had so derided. He sprinted to the water's edge where the Coral Sea lapped weakly against the wet sand, stripped off his $375 pair of sunglasses, and manfully flung them aside. He charged into the water, splashing up great heroic sprays from his feet as he raced through the shallows. He advanced as deep as his knees. Then he suddenly stopped.

Would it sting him, too?

She was about 50 feet away, in water that was probably just over her head. He could see her struggling to put her feet down to the sandy bottom. He stared hard at the water around her. He couldn't see anything unusual. How long were its tentacles? He had no idea. Where was it? It could be anywhere. She was looking up at him with the wide, panicked eyes of a wounded, trapped animal that knows it doesn't have much time. She was trying to call out to him, but her words were garbled with the seawater splashing into her mouth as she clawed and thrashed, as if pushing away invisible tentacles in the water around her.

"Where is it?" he shouted. "Where is it?"

It was as if Gil, standing paralyzed in knee-deep water near shore, had also been injected with the jellyfish's powerful, stunning venom. He looked up and down the beach for something, anything, that might help—a person, a boat, a long stick, a board, a rope. Nothing. They'd chosen this beach precisely because no one was here. What did he have in the car? Nothing useful. He looked back to her, thrashing in the water.

"Can you swim?" he shouted. "Swim to me!"

She raised an arm helplessly toward him, and he could see the red weals already spiraling along it in raw whiplashlike marks as she gasped and gagged on the seawater.

He realized that she was going to die unless he did something. But what? He could run to the car and drive for help. How long could she last? Five minutes? It would take at least an hour to get to town and back. He could see them charging down the beach, and there'd she be, washed up on shore, dead. And what would happen then? Would there be a coroner's inquest, or the Australian equivalent? Would he be exonerated if she died? Could he somehow be held as an accessory in her death?

No one knows exactly what's in the venom of *Chironex fleckeri* that makes it so potent. Research has been stymied partly because *C. fleckeri*'s venom is thermolabile—meaning it becomes unstable when warmed to 131 degrees or above, making it difficult to handle or analyze without chemically altering it. Researchers believe the venom, made up of proteinlike substances, contains three major components: one toxin that causes dermatonecrotic damage (damage and death to the skin), another that affects the blood, and a third that works, potentially fatally, on the heart and other organs. The incredible skin pain, some have speculated, may be due in part to a compound called 5-hydroxytryptamine, a pain producer that is found in the tentacles of many types of jellyfish. This compound triggers the release of histamine from the victim's body. While it is also one of the irritating chemicals in bee stings and nettles, histamine plays a helpful role in the body's immune system by opening blood vessels—thus causing inflammation around a wound—in order to speed immune agents to the injury.

But the pain caused by histamine inflammation is relatively mild compared to the incredible agony a *C. fleckeri* victim experiences. There is some other substance at work as well. Not much is known about how the venom affects blood cells; what is becoming clearer, however, is how the venom can affect the heart.

It has been estimated that *C. fleckeri* venom enters the circulatory system of a healthy person within twenty seconds of the sting, as the tiny tubules inject thousands of tiny doses of venom directly into the capillaries just beneath the skin. This is unlike snakebite, where the fangs leave a few large deposits of venom in the tissue, and this venom is slowly absorbed through the body, the full effects taking hours. In both cases, the fight-or-flight response—which might be useful if one were being attacked by a saltwater crocodile and needed the response's physiological boost to extract one's leg from the croc's mouth, as one near-victim managed to do—can work against the victim of jellyfish stings or snakebite.

Mary's fight-or-flight panic had sent surges of adrenaline through her body, boosting her heart rate from 80 beats per minute to its maximum of 170 beats. Her struggling contracted her muscle tissues, which demanded oxygen, triggering a flood of blood from her heart and lungs to her thrashing muscles and back again. The effect is what physiologists call the "muscle pump," and it helped carry the venom from the capillaries beneath her skin to her heart.

The venom quickly concentrated in her heart. Something in the venom—a cardiotoxin, researchers call it—began to wreak havoc with the heart's electrical system. In its normal resting state, a muscle cell of the heart holds a negative electrical charge inside it. This is because it contains negative ions—certain types of atoms that are missing electrons. When an electrical message passes from one muscle cell to another, channels open in the cell membranes, allowing negative ions to flow out of the cell and positive ions from outside the cell to flow into it. For an instant the charge inside the cell becomes slightly positive. This impulse is passed on to the next cell and the next, like a wave traveling along the muscle fiber, causing it to contract, and as it passes the positive ions move back outside the cell again, and the cell's charge returns to its original negative state.

But the cardiotoxin in *C. fleckeri* venom, according to experiments conducted on the heart tissues of rats, apparently triggers a big upsurge in the number of calcium ions with a positive charge entering the heart's muscle cells. An excess of calcium ions inside these cells is well known to cause spasms in the carefully timed sequential contractions of the heart muscle, a bit like throwing a bucket of water onto the circuitry of a spinning electrical motor.

Suddenly Mary's heart lost its rhythm and jumped with irregular, spasmodic contractions. The ventricles, the heart's two main chambers, normally start their contraction from the bottom point of the chambers and work smoothly up the powerful muscle walls to the top of the chamber, squeezing out the blood. But the jelly's venom triggered chaotic contractions that began somewhere in the middle of the muscle, and at an extremely rapid pace—240 beats per minute. Instead of a powerful, coordinated stroke pushing out 3 ounces of blood with every beat, it dribbled out less than one-twentieth that amount per beat. Her blood pressure plummeted. The blood flowing to her brain slowed to a trickle. The bright thrashing circle of sunlight and seawater and raw pain faded to a dim, lazy pool of twilight. *Keep your head up*, came the message from somewhere deep in that pool. *Keep your arms moving.*

It seemed to Gil she was losing consciousness. Her head dropped facedown in the water, bobbed up, dropped down again.

She wasn't screaming now. She floated facedown in the water, but the back of her head still protruded above the surface. Her lesioned arms still waved vaguely keeping her afloat.

"Mary! Mary!" he shouted, shuffling his feet a foot or two closer across the sandy bottom.

She gave no indication that she heard him.

Gil had never seen anyone die before, but it was perfectly clear to him that she was almost dead. He knew he'd replay this moment for the rest of his life, carrying this image of her, floating facedown, her arms still stirring dully, her hair gently fanning out on the surface, while he stood paralyzed, his feet fixed to the sandy bottom in knee-deep water and simply . . . watched. It wasn't a charging lion he was confronting, or a marauding bear, or even a hungry saltwater crocodile. It was a lowly jellyfish! One that just happened to be swimming by! People would laugh—the man who watched his wife die because he couldn't take on a jellyfish. He felt his lungs hyperventilating and his heart pounding. In Gil's panicked state—torn between fighting the jellyfish and fleeing from it—his heart rate had jumped nearly to its maximum of 150 beats per minute. He couldn't bear it—the idea of living the rest of his life thinking of himself as an inch-at-a-time, always-weigh-the-consequences tax lawyer who watched his wife die because he was afraid to help. Because he'd been so cautious, so afraid to die, he'd also been afraid to live. And now, because of his caution, his wife was about to die, too.

He moved one foot forward across the sandy bottom. Then he moved the other. Then the first again. Suddenly he was thrashing into the water toward her, no longer caring what happened to him, what happened to either of them, as long as he *acted*, churning his way out, swinging his arms back and forth for balance. Thigh deep . . . waist deep . . . chest deep. She lay floating only 10 feet away. . . . 6 feet. He stopped. He shuffled two steps closer. He stretched out his right arm, reaching up over the surface of the water, above any stray tentacles. He touched her left hand, feeling the sticky softness of the tentacles that were wrapped around it. He pulled away. Oddly, he felt no stinging. Were they somehow spent?

They were not spent; he felt no sting because the tiny venom-filled tubes launched by the nematocysts didn't have enough power to penetrate the thick skin of his palm. The hair on the back of his hand also acted as a barrier to their penetration. Women and children, who have more hairless, tender skin and a smaller body mass than men, are for this reason more susceptible to *C. fleckeri* stings. Even a layer of panty hose can repel jellyfish stings, as discovered by surfers and Australian life guards, who pull pantyhose over their legs and other pairs over their torsos, cutting holes in the seats to poke their heads out, when swimming in jellyfish-infested waters. Sea turtles are equipped with some sort of cast-iron gullet that allows them to chase down and gobble up with great relish large quantities of these jellyfish.

Gil reached out again. He took a grip on her left wrist, between tentacles. Adrenalin pumping, he dragged her 122 pounds splashing through the shallow water and pulled her up onto the dry sand.

He dropped to his knees beside her. His total focus was on how to revive her, an intensity of concentration he'd never before experienced, a kind of crystalline clarity unobstructed by ceaseless whispers of caution. She lay facedown. Loops and S-shapes of tentacle torn by her thrashing from the jelly's delicate body adhered to her torso, etching the spiraling purplish brown lesions on her skin. If she survived, the tissue could die, and she might be scarred for life with the whorled markings of *C. fleckeri* tentacles, as if tattooed by a heap of rope.

Gil remembered the sign posted at the beach near town. "Flood sting with vinegar" was the first instruction for treating severe box jellyfish stings. Gil had noticed the jugs of vinegar placed at regular intervals along that beach. Dousing the adhering tentacles with vinegar prevents the nematocysts from firing further. But Gil had no vinegar. Instead, he ran to his beach towel that bore the resort logo in big blue letters, snatched it up, and, running back to her, wrapped it around his right hand. Working his fingers through the fabric like a surgical glove, he quickly plucked the tentacle fragments from her skin, flinging them to the warm sand behind him.

About 12 feet of tentacle had made contact with her skin. The severity of a *C. fleckeri* sting depends largely on how much tentacle has made close contact with the victim; about 6 feet is considered roughly the minimum for a lethal dose in an adult human. By some estimates, a full-grown *C. fleckeri* contains enough venom in all its tentacles combined to kill between ten and twenty adult humans.

Gil rolled her over and quickly pulled the tentacle fragments from her midriff. As a precaution, he had taken a course in cardio-pulmonary resuscitation years earlier and a refresher course before their trip to the Caribbean, nervous about the medical facilities—or lack of them—in the islands. He placed one hand under her neck and tilted back her head to make sure her breathing passages were open, then watched her sand-covered abdomen and chest for a faint rise and fall of respiration. Nothing. He put his face close to her mouth and nose to feel a flow of air. None.

The sign had instructed: "If breathing stops, give artificial respiration. Give closed-chest massage if heart stops."

He could feel his own heart pounding, his hands trembling, his breath coming in short gasps. *Concentrate,* he told himself. *Remember the CPR sequence.*

He leaned down over her, opened her mouth with one hand, pinched closed her nostrils with the other, placed his open mouth over hers, exhaled once, turned aside, took a breath, and exhaled into her mouth again. He watched her chest. It rose slightly as the air went in, then fell. This meant that, although she needed artificial respiration, her air passages weren't blocked. But what about her heart? He placed his index and middle finger on her throat and slid them to the side, slipping them into the slot between trachea and neck muscles to feel for the pulse in the carotid artery that supplies blood to the brain. He held his sand-gritted, trembling fingers as still as he could. Nothing. Her heart was still in wild and chaotic spasms at over 200 beats per minute. But it is not possible to feel a pulse in the carotid artery with one's fingers if the victim's systolic blood pressure—the pressure during the heart's contraction phase— is less than 80, compared to a normal of about 120. With her heart in spastic contractions from the *C. fleckeri* venom, Mary's systolic blood pressure now stood at a mere 40. And her diastolic pressure—the pressure during the heart's expansion phase, usually the weaker of the two—was zero.

Gil had to become both her heart and her lungs, at least long enough for her own physiology to recover from the venom's shock and regain some of its equilibrium, and he knew this was not a simple task for one person alone to perform. He'd have to work extremely quickly. Already the anxious sweat spilled down his forehead, from his armpits, along his forearms. Kneeling beside her sprawled form, he laid his right hand on top of his left and placed them on her breastbone (careful to avoid its lower tip, or xiphoid, which could snap off under too much pressure), stiffened his elbows, and shoved down hard, compressing her sternum about 2 inches, then released and shoved again. He tried to keep a quick, steady rhythm of 80 thrusts per minute, although he would have aimed for a slower pace of 60 thrusts per minute if he'd had help and didn't have to stop periodically to ventilate her lungs himself. With his tax lawyer's mind for numbers, he easily recalled the key ratio for a person performing CPR alone—15:2. After the first 15 thrusts he leaned down, pinched closed her nostrils again, and gave her two quick breaths, then resumed another set of 15 compressions. *One-and-two-and-three-and-four* . . .

Though the air he breathed into her was expired from his own lungs, it still contained plenty of oxygen. With each downward thrust of her sternum, Gil squeezed her heart's right and left ventricles, one pushing blood to her lungs, where it could absorb the oxygen, the other sending the oxygenated blood coursing throughout her body. His rapid, steady compressions kept blood moving up her carotid arteries and into her brain at one-third its normal rate of flow—not a lot, but enough to deliver the crucial supply of oxygen to the brain's starving tissues.

Thirty seconds passed. Gil leaned into his chest compressions, forcing the blood through her heart's chambers. He didn't dare look up, didn't dare speak to her, didn't dare urge her back to life, didn't dare utter a prayer, so intent was he on maintaining the quick, steady *compressing, compressing, compressing . . . breathe . . . breathe . . . compressing, compressing . . .*

A minute passed. He paused for an instant, panting to catch his breath, to feel with his shaking fingertips for a pulse in her carotid artery again. Still nothing. He went back to his intense, steady rhythm.

The pumping action he forced on her heart now began to wash away the venom that had concentrated there, diluting the venom-infused blood that had come from her surface capillaries and dispersing it throughout her body. The cells of her heart muscle began to return to their normal complement of calcium ions, and the tiny holes that had opened wide in the cell membranes allowing ions to pass too easily in and out of the cell started to close.

Another forty seconds passed. Sweat flying from Gil's face and arms fell onto Mary's sand-crusted face. He didn't know how long he could keep up this pace.

As the venom moved away, her heart slowly regained its coordinated rhythm. Inside the roof of the right atrium, one of the two small chambers that sit above the ventricles, is embedded the heart's own pacemaker—a bundle of special muscle fibers called the sino-atrial node that, like a metronome, generates its own impulses. As it fired, it sent electrical impulses traveling through her heart muscle like a wave, ordering the firings of the exact proper muscle tissue in the exact proper sequence.

The second minute passed—he'd now done eight sets of 15 thrusts. Gil stopped again to probe for a pulse in her carotid artery. He felt . . . something. But what? Was it his own thumping heart, pushing blood out into his fingertips? Sand coated his fingers, their tips half numbed from the pressure of giving the compressions, and he was panting and trembling and couldn't see through the sweat dripping in his eyes. He swiped it away with a forearm, felt again. Was it his pulse, or hers, or his imagination? Whatever it was, he had to do something—fast. He needed someone to help him— someone with calm, clean, steady fingers.

He placed his hands on her sternum again and resumed the frenzied pace of the chest compressions, punctuated by the breathing. Another four sets of 15 thrusts—another minute. He was panting so hard now, he had to pause to catch his breath. It seemed that he'd been working over her frantically for a very long time. Was this futile? Had he got to her too late? Had he stood in the shallows too long while her life slid away? He felt again for her carotid artery. He held himself over her as quietly as he could, trying to understand what he sensed in his fingertips. It was then that he first heard a small gasp. He looked at her chest. There was another small gasp. He saw her rib cage fall slightly.

He let go of her neck and thrust his ear to her chest. He could hear something. He pressed his ear closer against the gritty skin. *Lub-dub . . . lub-dub . . . lub-dub . . . lub-dub*—it was the sound of her heart's valves opening and closing as they should, causing the blood to vibrate the heart's walls and vessels.

"Keep going, Mary!" he shouted. "Keep going!"

He lifted his head from her chest. Now her breath came in small but regular gasps—the half breaths known as agonal breathing.

"That's right, breathe, Mary, breathe!"

He watched closely as her chest moved up and down more firmly. Had it actually worked? That he'd breathed life back into her?

Her limbs began to stir, brushing against the dry sand as if she remembered, deep in her subconscious, that she was supposed to swim. Her head started to loll back and forth on the sand. Gil quickly brushed the crust of sand from her face. Her eyes were still closed. He needed a plan—and quickly. He decided that the moment her eyelids opened, he'd pick her up in his arms and haul her over the soft, boggy sand to the Jeep under the palms, prop her in the passenger seat, jounce over the rutted beach road to the highway, and race into town. They'd arrive in twenty minutes, maybe a little more, and the hospital in town would surely have a ready supply of the box jellyfish antivenin. They'd put her on a respirator if she still needed it, and give her medicine for the pain that would surely return.

What would happen then? he wondered. What would they say to each other? Would she in some way be a different person, not pushing him all the time? And would he be different, not hanging back? Could a brush against this odd, drifting, tentacled creature that had no brain—no sense of good nor evil, of right nor wrong, but simply was—change a life, two lives, a marriage?

He jumped up. He ran to the pile of clothes they'd shed on the beach and grabbed his daypack with the car keys in it. He knelt again beside her.

"Come on, Mary, come on!"

He allowed himself to wish now, to hope. She could relapse— *Chironex fleckeri* victims sometimes show a brief improvement and rising blood pressure before suddenly expiring—but Gil knew that he'd done for her everything that was humanly possible for him to do, far more than he would have ever guessed he could. As Mary's eyelids slowly opened, a surge of relief and gratitude welled up inside him, and tears began to sting his eyes.

CHAPTER NINE

BUBBLING FROM

THE BOTTOM UP

THE BENDS

Slightly out of breath from racing through the cobblestone streets to arrive at the appointed hour, Robert—or Roberto, as he'd been known for these past nine months in Spain—strode up broad stone steps to the massive black door of an old merchant's house at 49 Calle San Jose in the ancient port city of Cádiz. Robert knew the seaport's history well. Claiming to be the oldest continuously occupied city in Western Europe, Cádiz's heavily fortified harbor was first constructed thirty centuries ago by Phoenician traders on a rock-and-sand spit just outside the Straits of Gibraltar. Cádiz had served as the first port of call in the Old World for many of the treasure fleets arriving from Spain's vast American colonies. It was this last fact that interested Robert most.

Wiping his sweaty cheeks and forehead with a handkerchief, he composed his face into what he thought was a dignified but friendly expression. He then pressed the buzzer mounted on the stone jamb.

"*Qué?*" a voice said through the speaker.

He gave his name. There was no response, only the harsh buzz of the electronic lock pulling back.

He pushed open the heavy door. He'd been in old Cádiz merchant houses before—many had been cut up into apartments—but he'd never seen one so beautifully preserved. He followed the tunnel of a barrel-vaulted stone foyer until it suddenly jumped open in an atrium that reached up a full five stories to the sky. A tall palm stood in its center, and its marble floor—black and white squares like a chessboard—had been hollowed in places by centuries of boot traffic striding across it to the in-house merchant offices that once occupied the ground floor and mezzanine.

"*Aquí!*" a voice called, echoing through the atrium. "*Venga!*"

Robert looked up. A young woman in a black skirt and white blouse stood on the third-floor balcony, its filigreed wrought-iron railing encircling the atrium like the balconies above and below it, each leading off into several doors.

Robert started up the marble stairs. On the landings he passed antique Mudejar chests inlaid with ivory arabesques created by Moorish craftsmen who lingered in Spain after the Christians had reconquered it starting in the eleventh century. There were chairs upholstered in embroidered leather dating from the sixteenth century. Everywhere Robert looked he saw expensive relics from Spain's distant and glorious past. *This has got to be the place,* he thought.

At the top of the stairs, the young woman led him along the balcony into a sitting room. There, in a straight-backed armchair with her hands carefully resting on the arms, sat an elegant old woman. A white Andalusian shawl, laced like the web of a spider, was draped over the shoulders of her black dress. Her white hair was pulled back so tightly, it stretched the skin against her skull.

She lifted a hand to gesture to another straight-backed chair, this one upholstered in red velvet with spiraling Moorish-style legs. Robert sat down. The chair creaked and popped under his weight. He felt the sweat trickle down his cheeks.

"You're not the first, you know," the woman abruptly began.

"Someone has been here before me?"

"Not since my time, " she replied slowly, her lips moving carefully. "I only know what my grandmother told me when I was a girl. She heard it from her mother. Someone went to look for the ship.

They say that no one will ever reach it because the ship lies *muy hondo.* Very deep."

"But that was many, many years ago," Robert said. "Now maybe it is easier to find with modern diving equipment. It's possible to go very deep."

"Quizá," was all the woman said in reply. Perhaps.

As they sat, Robert sweating in the creaking, straight-backed chair, the young woman who had led him in carried in a tray bearing a silver coffeepot. Robert instantly noticed the dented heaviness of the silver and the rich design, an encircling band of squared, intertwined frogs, almost Mesoamerican in spirit, of the type Aztec craftsmen made under Spanish rule. The two of them watched as she poured out with slender white hands a thick, black stream of coffee into two cups.

"This is my granddaughter," said the old woman.

For the first time, Robert, so transfixed by the house's furnishings, closely noticed the young woman's face. In a quiet way, she was beautiful. She had a finely sculpted nose, a strong but delicate jaw line, and jet black hair piled on her head that accentuated her white skin. Her eyes were dark and deep and wide when she looked up shyly at Robert to ask him if he'd like anything in his coffee, and her lips were carefully painted a blush of red.

"A little sugar, if you have it, *gracias,*" Robert said.

He watched her dextrous fingers, the long nails painted red, measure out a spoonful and a half of sugar, mix it into his coffee, and hand him the cup and saucer with a slight bow of the head. Then she straightened up and left the room.

"Why do you think I can help you?" the old woman asked.

"Please, *señora,*" Robert began. "I've been to the archives in Sevilla. For months I've been reading the old manuscripts and the merchant bills of lading. I know about the storm in 1605 and the loss of the ships of that year's fleet. As I explained in my letter to you, I know how much your ancestor lost. I know the ships and his lost wealth have never been recovered."

"Why do you want to search for *oro y plata*?" she said. "Gold and silver have given blood and trouble enough to the world."

"I collect things, " Robert replied, suddenly aware that this was not much of an explanation.

The ancient Scandinavians believed that a malevolent giant by the name of Aegir lived under the sea, married to a lascivious and greedy wife named Ran. To satisfy her lusts, Ran stirred up the sea with tempests, capsized the Scandinavian ships, and caught the drowning sailors in her nets as they disappeared beneath the waves. If the sailors gave her a tribute of gold, she rewarded them with an eternal place at her undersea banquet table and in her watery bed.

Ran, however, was only one of many potential lovers for a drowning sailor. Seductive and treacherous "women of the sea" occur in the myths of maritime cultures around the world. In the ancient Mediterranean, the Sirens, beautiful but deadly nymphs, used their sweet, hypnotic songs to lure sailors toward the rocks where their ships would founder. For the Irish and Scottish, the White Ladies made their home beneath the waves, and the Slavs believed the sea hid the beautiful drowned girls known as Rousalki. The Eskimos tell the story of the luscious but handless Nerrivik—whose name translates literally as "Food Dish"—who lives at the bottom of the sea and whose long hair must be combed by shamans to release the bountiful fish and sea creatures that are tangled in her locks.

These female myths represent the two faces of the sea—calm and alluring, tempestuous and deadly—with which sailors have lived and died for centuries. Deep beneath the waves the sea does not show its tempestuous side. It offers instead a peacefulness and aquamarine beauty and silent mystery that are just as seductive as the sparkling blue surface and perhaps even more deadly. Divers, like ancient and modern sailors, have many terms, medical and other-

wise, to describe the physiological consequences of this undersea se-
ductiveness. Perhaps the most apt was coined by the pioneering—
and poetic—French divers describing what happens to those breathing
compressed air who descend too far into the serene depths: "rapture
of the deep."

The human body is surprisingly adaptable to the depths, at
least on the way down. All humans, in different ways, thrive on pressure. Not just pressure from work or money or romance, but actual
pounds-per-square-inch pressure. As a species, humans have adapted,
more or less, to live contentedly with 14.7 pounds per square inch,
the air pressure at sea level. The 14.7 pounds—the equivalent of an
average frozen turkey perched on every square inch of one's body—
represents the weight of a very tall 1-inch-by-1-inch column of air
that extends from sea level many miles up to the farthest edge of the
earth's atmosphere. This pressure pushes on us equally from outside
as well as from inside—inside our lungs, our throats, and elsewhere—
and is the key reason why we don't notice the groaning weight of all
those frozen turkeys piled on top and pressing around us.

Water, however, weighs vastly more than air. A 1-inch-by-
1-inch column of seawater would only need to be 33 feet tall to weigh
the same amount—14.7 pounds—as that miles-tall column of air.
This difference in weight has profound consequences for humans
who leave the relative comfort of sea level and dive under the sea. If
a person sitting in a boat were to jump over the gunnel and swim
down to a depth of 33 feet, the pressure on his or her body would
double from the standard 14.7 pounds at the surface to 29.4 pounds
per square inch, or two frozen turkeys. This is known as two atmospheres of pressure. At 66 feet in depth, the pressure would triple to
three atmospheres, or 44.1 pounds per square inch (psi); at 99 feet, it
would quadruple to four atmospheres or 58.8 psi; and at 132 feet,
about the recommended limit for recreational divers using scuba
gear, the pressure would be five atmospheres or 73.5 psi. At this
depth the average-sized human male experiences a total of about
100 tons or 13,400 frozen turkeys pressing over his body surface.

Along with the obvious need to breathe, divers for thousands of years have struggled with these tremendous pressures of the deep. Greek sponge divers of Aristotle's time knotted ropes around their waists, packed their ear canals with oil-soaked sponges to help equalize the pressure stabbing at their eardrums, took a curved knife in one hand and a heavy stone in the other, and plummeted to the bottom, relying on their compatriots above to pull them—sometimes dead—to the surface after they'd gathered the harvest. The ancient Arab pearl divers pierced their eardrums to relieve the pressure of the depths. Today's breath-holding pearl divers of the Tuamotu Archipelago, northeast of Tahiti, dive to over 100 feet many times a day; as a result, they frequently suffer from the sickness they call *taravana*—"fall crazily"—whose symptoms include dizziness and nausea and, in severe cases, paralysis and death.

The perfection of the diving suit in the mid-nineteenth century, while allowing divers to go deeper and stay down longer than those who simply held their breath, only aggravated these pressure-related maladies. In the perfected design, compressed air was pumped from the surface down through a hose to the hard-helmeted diver. (Common sense says a diver only needs to poke a tube up to the surface and suck ordinary air through it, but the water pressure on the diver's chest makes it impossible to inhale using this method in water any deeper than snorkeling depth—about 1 to 2 feet.) Those Aegean Sea sponge divers abandoned the methods Aristotle had described and in 1867 adopted this new diving suit, trying to maximize their profits by staying at the bottom for long periods of time, then ascending quickly. Of twenty-four divers using the new suits, ten of them died.

As the Industrial Age surged forward, other inventions allowed workers to breathe deep beneath the surface. French engineers devised a new method to build bridge pilings by placing large metal cylinders on the river bottoms, pumping these "caissons" full of compressed air, and placing workers inside to construct the bridge footings. Strange new maladies, however, began to appear. During construction of the Brooklyn Bridge in the late nineteenth century, workers emerging from the caissons walked about bent stiffly forward at the hips due to the sudden onset of joint pain. Their fellow workers thought the gait resembled the fashionable "Grecian bend" of the tightly corseted New York society women. The name "the bends" stuck. During World War II, French gas engineer Emile Gagnan and a soon-to-be-famous diver named Jacques-Yves Cousteau invented the self-contained underwater breathing apparatus, or scuba—a system that allowed divers to dispense with the cumbersome suit and hose and carry their air supply in tanks on their backs. The genius of the Gagnan-Cousteau equipment was a small disklike device on the mouthpiece called a demand regulator that, using a diaphragm to sense the water pressure, fed compressed air from the tanks to the diver at the exact same pressure as the water pushing on his lungs from outside. With the pressure exactly equalized inside and outside the chest walls, breathing far beneath the surface was effortless. But divers using the gear, it turned out, were subject to the same "bends" that caisson workers suffered. And there were still other maladies that divers came to know as "the chokes," "the itches," "the staggers," "the narcs," "the niggles," and "the creeps," having to do, variously, with uncontrollable coughing upon surfacing, itching of the skin, dizziness and nausea, and odd but fleeting aches.

Starting with autopsies on Greek sponge divers, researchers continued to try to understand what great pressure does to the human body, especially divers breathing compressed air. They discovered that the pressures of the depths drive nitrogen contained in the compressed air into the diver's blood and tissues. A breath of air is, after all, four-fifths nitrogen and only one-fifth oxygen. At depths over 100 feet, the nitrogen absorbed by brain tissues can anesthetize the diver like laughing gas. This is the French "rapture of the deep"—also known as nitrogen narcosis or "the narcs." If the diver ascends slowly from these depths, the nitrogen seeps out of the blood and tissues on the way to the surface and is harmlessly expelled by the lungs. A too-rapid ascent, however, causes the pressurized nitrogen to fizz inside the tissues and blood like the gas bubbles in a just-opened soda bottle. These bubbles cause the bends. Its most common symptom consists of pain in the knees, shoulders, and elbows—"limb bends" or "joint bends." But the bubbles or related symptoms can also occur in many other places such as the skin, and, much more seriously, in the spinal cord and brain. When this occurs, the bends can result in paralysis or death.

At the turn of the century the eminent British physiologist J. S. Haldane, who worked with bends-afflicted divers and depressurized goats, concluded that the human body generally can tolerate a two-to-one reduction in pressure without suffering the symptoms of decompression illness. This would allow a diver to ascend from two atmospheres to one (33 feet to the surface) or from four atmospheres to two atmospheres (99 feet to 33 feet) without fear of the bends. But, at least according to Haldane's original model, ascending above the depths specified by this ratio without pausing to "decompress" would place the diver in danger. Based on this work, Haldane developed a set of schedules—much refined by later generations of divers—that they could use to plan their dives and avoid the bends.

If the diver does suffer from a case of the bends, he or she must be recompressed—recapping the soda pop bottle, so to speak—by returning below the surface and resuming the ascent in stages to let the nitrogen reabsorb in the tissues and then slowly seep out. This treatment is a long and difficult process to accomplish underwater. A better alternative is to place the diver in a recompression chamber and pump it full of compressed air to mimic the high water pressures undersea. This act accomplishes the same thing—allowing the nitrogen to seep out by slowly decreasing the pressure. One problem with this treatment, however, is that recompression chambers are often located far from the remote reefs and islands that divers seek out, and that flying in an airplane to reach a chamber can create more bubbles in the stricken diver's body as the plane ascends and cabin air pressure drops. The evacuation plane must fly low—below 1,000 feet.

The bends, however, is not the most fatal of the pressure-related maladies grouped under the term *dysbarism*. Drowning still remains the leading cause of death among divers. The second most common death is a form of dysbarism that occurs when, due to panic or haste, the diver doesn't breathe out regularly as he or she swims to the surface. The compressed air in the lungs expands with the lessening of the water pressure like an inflating balloon. If the diver keeps holding his or her breath, the lung tissues will rupture. Through the ruptured lung tissues, the air in the lungs then enters directly into the bloodstream, and coarse bubbles foam through the heart and into the brain. The result is cataclysmic.

To dysbarism researchers and doctors, this syndrome goes by the acronyms POPS and AGE—pulmonary overpressurization syndrome accompanied by arterial gas embolism. Divers, however, partial to their own nomenclature, know it by a term that much more graphically illustrates its pathology: "burst lung."

"Nothing down there for you to see, mon. All the pretty fish are over there."

The small wooden boat rocked on the blue swells just off the Caribbean reef, its outboard motor idling with a rhythmic gurgling sound. Robert was pointing out to the fisherman, Felix, the exact spot he wanted to dive, a deep channel that cut through the reef's paler greenish waters like a deep, thick vein of the sea's dark-blue blood. Felix, however, was gesturing in the other direction, half a mile distant, to the better place where the pretty reef fish lived. Ignoring him, Robert again consulted his chart and then the digital readout on his handheld GPS—the global positioning device that read satellite signals to give his exact position on the face of the globe. He jabbed his finger to where the deep blue channel cut through the foaming white crest of the reef.

"Right there."

"Nothing but sand down there."

"That's the place."

"Why do you want to go to that place?"

"Because that's where I want to go," Robert insisted. "Because I'm paying you to go there."

"Okay, mon. I'll take you there if you want to go, but I tell you for certain you see only naked sand down there in the deep water."

Two days earlier, Robert had claimed a window seat on the wide-bodied jet flying over the Atlantic Ocean between Madrid and Mexico City. He'd left back in Cádiz the old woman in the straight-backed chair and her granddaughter with the beautiful dark eyes and quiet manner. Robert remained motionless with his forehead pressed against the cold, clear plastic and felt the high hum of the engines in his skull while each of his slow, steady exhales lightly fogged a patch of window near his mouth. From 33,000 feet up, the Atlantic appeared so deep a shade of blue, it looked almost black. Focusing hard with his tired eyes, Robert could just make out the tiny corrugations of the swells and the chalky white flecks where the biggest swells had crested and tumbled into long streaks of foam. It must be rough as hell down there, Robert thought. He wondered how long it would be before he was bobbing among those same swells, or rather, diving under them. His ears suddenly popped with the decreased cabin pressure, finally clearing.

Robert had always been a collector of things. As a boy, he dug through small dark spaces—closets, basements, into the earth itself—seeking marbles or pennies, beetles or worms, old bottles or square nails. He assembled his treasures into collections, neatly gluing or pinning the objects to pieces of labeled cardboard, then carefully locking them away in a special plywood box in his bedroom. The collections were his secret world, and Robert protected them fiercely, guarding the relics he so eagerly sought out.

Staring out the window at the midnight blue Atlantic 6 miles below, wondering what might lie under its opaque surface, it occurred to Robert that all humans were genetically wired for the hunt, the perpetual search for bigger game to kill or fatter roots to dig or sweeter fruit to pick, but that some had chosen—by chance or by personality or by necessity—to refine and develop that innate skill to its fullest potential. Robert realized that he would soon put that highly developed skill of his to its most difficult test ever.

Now, two days later, Robert was eager for the dive to begin. Felix reached back, flipped the outboard's gear lever into forward, and twisted the throttle. The heavy wooden boat heaved over the swells toward the broad, deep blue stripe of the channel. Just short of it, still over the reef's submerged shelf in about 40 feet of aquamarine water, Felix cut the engine and flung the anchor overboard.

"Now you go," Felix said. "Over the side and swim to the deep spot." In silence but for the sloshing of the swells up the sides of the boat's hull and the clanking of his air tanks, Robert began to prepare his equipment. He'd hired Felix in a fishing village on a small island off the coast, surprised at the fisherman's willingness to take him out on the water as soon as he wanted.

"Wherever you want to go," he'd told Robert, "I'll take you there."

Felix now watched Robert's careful preparations. Robert knew how to dive from his days as a collector, combing lake bottoms to hunt for old bottles that had been discarded there decades before. Robert strapped his air tanks to his BC, or buoyancy control—a vestlike device into which the diver could inject or release air to maintain neutral buoyancy while underwater. Robert spun his tank valve full open, then closed the knob back a quarter turn. The air pressure gauge jumped to 3,000 pounds per square inch, indicating the air pressure inside the tanks. Each of the two tanks contained 80 cubic feet of air squeezed to a fraction of that volume, which would sustain him beneath the surface. He punched in the diaphragm of the regulator mounted to the mouthpiece and checked the loud hiss of escaping air. He then released it, inserted the mouthpiece, and inhaled—once, twice, three times. The air flowed easily into his lungs. He examined his spare regulator, or "octopus," attached to its own hose should his main regulator fail. He consulted the all-important dive-planning tables. At a depth of 90 feet—as deep as the bottom of the channel, according to the charts—he could spend a maximum of twenty-one minutes. Any deeper or any longer and he would have to make stops on the ascent, spending several minutes at various depths while the nitrogen slowly seeped out of his tissues.

Though Robert now dove less than he once did, he intimately knew the safety checks. He meticulously followed all of them except one: *Never dive alone.*

"Okay, are you ready, mon?" Felix asked once Robert had slipped into the shoulder straps of the tanks, fastened its belts, and pulled on gloves, fins, and face mask. Propping himself on the gunnel of the dory, Robert fitted in the mouthpiece and sucked on the regulator, one hand holding his mask tightly against his face. He nodded to Felix.

"Okay, one, two, three, over you go."

With a little nudge from Felix, Robert tumbled backward off the gunnel into the swells in a gentle splash.

5 feet (1.2 atmospheres of pressure or 17.6 psi): A swirl of bubbles swept across his mask, and warm tropical water trickled into his ears and wetsuit. The vigorous rocking of the boat softened to the gentle up-and-down motion of the sea just beneath the swells, comforting and familiar, as if the gentle watery lifting and sinking awakened some primordial memory of the womb. The harsh sound of wind and waves suddenly ceased, replaced by the underwater clicking and scraping of unseen fish along the reef.

Robert righted himself from his tumble. He hovered for a few moments, belly down, to orient himself. In his left hand he held a small instrument console connected by a slim hose to his tanks. One dial was his air pressure gauge, another his depth gauge, and a third his compass. He took a compass bearing to the northwest, flipped his fins, and slanted down toward the bottom.

Robert knew exactly the direction he wanted to go, thanks to the old woman and her granddaughter back in Cádiz.

"Look at this room," the old woman said as she waved a thin arm at the ivory-inlaid *vargueño* cabinet, the thick walnut table with the wrought-iron stays, the high-backed armchairs in which they sat, the wooden chests with the star-shaped patterns, even the heavy beams of the ceiling that Robert knew had been hewn from Cuba's forests and shipped to Cádiz centuries before. "What am I to do with all these old things now?"

"They are very beautiful," Robert said politely.

"They belonged to my family for generations," she said, "back to those who lost their precious belongings when the ships went down in the year 1605."

From the bills of lading preserved so carefully in the Sevilla archives, Robert knew exactly what her ancestor's cargo had included: gold rods in various sizes and weights that merchants such as he had ordered smelted to ship home their wealth; an emerald-studded gold crucifix, gold rosaries, and other items of jewelry fashioned by the skilled New World gold- and silversmiths; plus hundreds of ingots of silver that had been laid like brick ballast in the bottom of the hull.

"The loss was a very difficult one for our *negocios*—our business," the old woman went on. "So I was told as a small girl by my mother and grandmother. But you can see"—here she gestured again at the room, at herself—"my family survived the loss."

Robert noticed that between the fingers of her left hand she worked a rosary. It made a tiny clicking sound as its alternating beads of polished red coral bumped its well-worn beads of gold. She now appeared lost in thought, staring out the window. Beyond the wrought-iron balcony and over the rooftops across the street rose the dome of Cádiz's cathedral. It looked like a great yellow ceramic bowl—almost a dome that belonged on a Moorish mosque. Robert knew that the cathedral occupied the same site as the mosque from the Moorish occupation that had begun in A.D. 711 and that the mosque had sat atop the site of Cádiz's original Roman temple. This probably had been built over its three-thousand-year-old Phoenician temple, where the ancient traders had no doubt practiced their ritual sacrifices of animals and human infants to the god of storms, Baal, to the fertility goddess, Ishtar, and to others. Since Robert had arrived, archaeologists working in the bishop's patio had unearthed a gold ring embossed with a swimming dolphin—a fertility symbol— that dated from Phoenician times. The old woman stared silently at the dome, clicking her beads through her wrinkled fingers. It was as if that particular spot were the repository for the ancient gods as well as for her family memories.

Robert leaned slightly forward in his chair toward her. He wanted to appear interested but not too eager. As he leaned, his shirt hung loosely in front and he felt a runnel of sweat trickle over his belly.

"Did your ancestors know where the ship went down?" he asked carefully.

"There are many letters describing the disaster," she replied. "Descriptions of reefs near where the ship sunk. But as I have told you, it is down too deep."

Robert leaned farther forward in the Moorish chair.

"Do you know what happened to the letters?" he asked. "Do they still survive?"

"*Claro,*" she said, with another click of the rosary beads.

"Where are they?"

"I have them here," she said, picking up a small brass bell that sat on an end table beside her chair and shaking it, the tingling sound echoing out in the old house's marble atrium. "They are under my bed. My granddaughter will show you."

Robert then leaned forward in his chair so far that it swayed on the verge of collapse.

As he descended, Robert paused every few feet to check his instruments and clear his ears. He did this by thrusting his jaw forward and by swallowing, opening the 1.5-inch long eustachian tubes that served as air passages from his throat to the air-filled chamber of his middle ear. If the tubes, inflamed or blocked by a cold, remained closed, the increasing water pressure on the outside of his ears as he dove deeper would collapse the tympanic membrane—his eardrum—inward, finally rupturing it. But by opening his eustachian tubes, Robert allowed the compressed air in his throat to fill the middle-ear chamber at a pressure equal to that of the outside water, thus keeping the eardrum intact. Middle-ear barotrauma is diving's most common medical problem, but even in its most serious form—a rupture—the damage usually heals in a week or two.

33 feet (2 atmospheres or 29.4 psi): A few feet below him, Robert spotted Felix's anchor lying amid patches of coral and sand. A large grouper drifted off to his right, where the reef sloped gently toward the deeper sea. The pressure on his body had now doubled since leaving Felix's boat. A balloon full of air—12 liters, for example—blown up in Felix's boat and brought to this depth would have now been compressed to half its size, or 6 liters, according to Boyle's law of gases. Likewise, each breath of air Robert's regulator fed him from his tanks was twice as compressed as a breath of air at the surface, meaning, all things being equal, that his air supply would last only half as long at this depth, and would last even less than that the deeper he dove.

Leaving the anchor, Robert followed his compass bearing of 340 degrees northwest across the reef's shelf toward the deeper blue cut of the channel. The swimming was easy. He stayed focused straight ahead. After a few minutes, the coral suddenly plunged steeply downward in an underwater wall.

55 feet (2.7 atmospheres or 39.7 psi): He paused at the dropoff, just able to make out the bluish, sandy floor of the channel far below. Brightly colored fish flitted past—black-and-white spotted drums, damselfish, indigo hamlets. Robert ignored them; he was focused on a far bigger catch. He checked his pressure gauge: There were 2,750 pounds per square inch of air remaining in his tanks. He checked his watch: seven minutes had elapsed since he left the surface. He felt good. His breathing came easily, although quickly. He felt a warm flush of excitement even under his wet suit. He loved the dramatic arc of the hunt: the first clues triggering that sense of excitement, followed by the patient reconnaissance of the terrain, then the anxious fever of closing in and finally the euphoria of discovery. As he angled down the reef's wall, his pulsing fins propelling him deeper, he was no longer a paunchy, lonely, middle-aged schoolteacher but a sleek, black-clad undersea creature. He was a shark that used his intellect instead of his sense of smell and movement to locate his prey.

72 feet (3.2 atmospheres or 47 psi): The reef's wall resembled a tropical flower garden tilted to the vertical and submerged beneath the sea. Robert swam down past the boulder coral, wire coral, black coral, and yellow tube sponges reaching out like fingers to strain tiny creatures from the seawater. His body now was experiencing a pressure of nearly 50 pounds per square inch. But he noticed physiological changes more subtle than profound: a pressing on his sinuses, the need to keep opening his eustachian tubes, and a slightly greater effort to breathe as the air he inhaled became more dense. According to Pascal's law, pressure applied to one part of a liquid is distributed evenly throughout that liquid. Robert, like all humans, was made up mostly of water; he felt no extra pressure on particular tissues because each cell in his body was pressed equally from all sides.

80 feet (3.4 atmospheres or 50 psi): Robert glanced again at his depth gauge, then down the reef wall to the bottom. He'd nearly reached the planned limit of his dive—90 feet—but the bottom, still

lying well below him, was farther than he'd thought. "Plan your dive," was one of the diver's mantras, "and dive your plan." He knew he should dive no deeper. But he was so close. He could feel the presence of the wrecked ship. He could sense the precious metals— the gold rods stamped with their purity, the silver ingots bound for the merchants and the royal treasury, the golden chains and scepters and emerald crucifixes. He knew he was in the right place. He couldn't bear the thought of leaving that collection, of wasting the dive, wasting the air in his tanks, wasting the time at the surface to let the nitrogen seep out of his system. Not when he was this close. No, this was his chance.

He'd stick to his plan, but he'd modify it slightly: he'd descend to 90 feet, maybe a bit farther, and swim along at that depth and try to scan the bottom.

99 feet (4 atmospheres or 59 psi): He now could see the bottom clearly, although he was still well above it. Flat and sandy, it was bathed in deep bluish hues—only the blues and greens of the light spectrum penetrated this deeply into the sea—and it was littered with broken pieces of coral. Hovering above it, he calculated his strategy: He'd remain at 99 feet only 15 minutes instead of the twenty-one minutes that, according to the dive-planning tables, he was allowed at 90 feet, and he'd make a brief "safety stop" on the way up. At this depth—four atmospheres—the pressure would squeeze a balloon to only one-fourth the volume it had been on the surface; his air supply would last only one-quarter the length of time. He'd have to watch closely the gauge that measured the air remaining in his tanks.

110 feet (4.3 atmospheres or 63.2 psi): Robert continued swimming on his 340-degree heading, rocking his head from side to side to scan the bottom, angling slightly deeper as if working along a line of tension between the upward pull of his dive plan and the downward pull of the wreck. He'd now dived so deep that a third law of the depths besides Boyle's law and Pascal's law came into play. Some divers call it Martini's law: Every additional 50 feet a diver descends

from the surface is the rough equivalent of consuming one martini on an empty stomach.

Nitrogen is one of several inert gases that dissolve easily in the body's fatty substances but become potent narcotics under high pressures, possibly by interfering with the signals jumping across the brain's nerve junctions. Divers who go too deep breathing compressed air risk the hallucinations, fantasies, and euphoria of nitrogen narcosis or "rapture of the deep." This state is famously illustrated by the apocryphal diver who offered his mouthpiece to a passing fish, or, in a confirmed case, the French diver who at 160 feet patted his nonexistent pockets looking for his cigarettes. The first subtle signs of nitrogen narcosis, depending on the individual, begin at about 100 feet. By 180 feet, experts agree that "no trust should be placed in human performance." By 300 feet, a special mixture of helium, oxygen, hydrogen, and nitrogen must be used to prevent nitrogen narcosis, as divers inhaling simple compressed air can black out and, of course, drown.

115 feet (4.5 atmospheres or 66.2 psi): Robert felt like laughing, joyous over this wild pursuit of his, this search for the lost wealth of the Indies. Like a childhood fantasy it seemed, this treasure hoard so nearby. How clever he'd been to track down the old merchant family in Spain! And once he'd finally found the old woman, it had been so easy! And her granddaughter was so beautiful!

The granddaughter had led Robert out of the sitting room and along the balcony circling the atrium. The heels of her slim, knee-high suede boots echoed on the black and white marble squares. Her black silk skirt fell in pleats nearly to the tops of the boots, but fitted snugly around her graceful hips, which flicked ever so slightly as she moved, as if keeping rhythm with the sway of her silver pen-

dant earrings. Robert could feel his breathing grow shallow with an-
ticipation. The granddaughter led him into a large bedchamber, a
room filled with still more antique furniture. The late-afternoon sunlight was falling through gauzy aquamarine curtains that hung in
windows set in thick white walls, the light bathing the room in watery, shifting hues of blue. A tall four-poster bed was draped with a
thick quilt of midnight blue.

She leaned down, pulled a small wooden trunk from beneath
the bed, opened it, and extracted a packet wrapped in old, stiff paper
and bound with a faded pink ribbon. She stood before him, holding
the old packet in both her hands, as if unsure what to do next.

"I'd like to make a proposal to you," he said, filling in the awkward silence. "If you and your grandmother share these letters with
me, I will share with the two of you whatever I find."

As he spoke, she stood with her brown eyes cast aside in shy-
ness. When he had finished, she looked out the window and over
the rooftops at the yellow-tiled dome of the cathedral that sheltered
the figure of Christ and all the gods of the earth and the sky, and especially of the sea, that had come before Him in this ancient port at
the gates to the Mediterranean.

She looked from the window back to Robert. He could feel the
sweat forming on his hands again. He subtly wiped them on the
thighs of his corduroy pants.

"My grandmother is an old woman and cares little for treasure,"
she said, finally fixing him with her deep brown eyes. "She will tell
you to give her share of it there, to *la catedral*." She gestured toward
the dome out the window. "So if you find it, you may do with the
treasure what you wish. But if you do find it, I ask that you set aside
a small share for me without telling my grandmother. Use that
share to take me away from here."

With that she stepped toward him and, peering into his eyes,
lifted up the package with both her hands, as if making an offering
to him.

Robert surveyed the bottom, kicking along, holding his instrument console in his left hand, keeping closely to 340 degrees. It couldn't be far now, he thought. And then he saw it. A spiky rod protruding from the sandy bottom. It lay below him by 15 or 20 feet. He glanced at his air pressure gauge—still 1,700 pounds left in his tanks—and without another instant's hesitation he put his head down, fins up, and dove for it. He was the predator coming to the glorious climax of the hunt, powerfully pumping through the water and closing in on his prey.

135 feet (5.1 atmospheres or 75 psi): He tried to shake it. It was wedged solidly in the sandy bottom. It felt like iron, he thought, encrusted with coral and tiny shells. Maybe the anchor shaft. Yes, it had to be the anchor shaft. Only the metal would remain. The ship's wooden parts couldn't survive four centuries of tiny marine animals feasting on them, unless buried in the sand. He looked around in the drab light. The objects looked clear and sharp, but their colors were washed out into bluish grays like an old black-and-white film shown late at night on a bad television set. About 50 feet away he spotted a low, sandy mound from which long tubes protruded like giant porcupine spikes. They were too uniform in shape for sponges. He swam quickly to the tubes. A fine sediment covered them. He brushed one with his neoprene glove. Small puffs of sediment floated up. He saw the dull glint of metal. Bronze. A cannon. An old Spanish cannon! He was in the right place! He looked past the mound. A deeper shelf dropped down another 30 feet or so. On it, below him, he spotted another mound. More wreckage!

165 feet (6 atmospheres or 88.2 psi): He kicked down along the incline. The pile was lumpy, sand-covered, unnatural-looking. His gloved hand penetrated easily into the sediment. Wriggling and work-

ing his fingers deeper, nearly to his elbow, he felt something solid.
He moved his fingers over the object. It was heavy and rectangular,
about the size of a brick. A silver ingot! He tugged at it. The accre-
tions of sand and marine life had cemented it to the surrounding objects. It didn't move. He felt about. More rectangular shapes. More
ingots! He tugged harder. Still he couldn't budge it. He extracted his
hand and began brushing the layers of sand from the pile, sending
up clouds of silt, along with great bursts of bubbles from his labored,
excited breathing.

Digging away with gloved hands toward the silver ingots, Robert thought of the Mayans and Aztecs and Incas whose sweat, blood,
gods were distilled here, in this mound of precious metal. Empires
rose and fell, and gold followed the shift in power, or was it the
power that followed the gold? Inevitably the two finally moved on—
from the Andes and the Yucatán to Spain, then to northern Europe,
then eventually to North America. This pile had been diverted from
the flow of power for four centuries, and now he'd claimed it. He
would ride it into that great historic flow of empire. His hunter's instincts had led him here, and now the treasure would change his life.
He wouldn't make the selfish mistakes that others had made over it.
He'd do good things with it. He'd share it, starting with the beautiful
granddaughter, who had asked him to take her away.

He'd now excavated a 2-foot-wide hole in the sand that covered
the pile. He let the sediment settle for a moment. As the water
cleared, he could see the black rectangles piled atop each other—
surely covered with the black tarnish of the silver.

He reached in to give another hard tug at one of the ingots,
sucking in a big breath of air and bracing himself for the effort. But
the breath only half filled his lungs. Suddenly there was no more air
coming through his mouthpiece.

He released his grip on the ingot and grabbed the instrument
console dangling at the end of its slender hose. At this depth he'd been
using up air six times faster than at surface pressures. The needle was

pegged all the way to zero, having crossed through the red warning zone that started just under 1,000 pounds per square inch of air in his tanks. It was very clear. There was no air left in the cylinders.

Looking up, he saw the last of his exhaled bubbles jiggling up toward the silvery surface far above. There were now sixteen stories of water above Robert. He was hovering at the very bottom of the channel, beside the collection of treasure.

It was the custom in Cádiz to kiss acquaintances of the opposite sex—even near-strangers—on both cheeks in departure. After she had given him the packet of old letters in the bedchamber, the granddaughter had reached up and kissed him on both cheeks. Then, in a grip that was surprisingly strong, she had pulled him toward her and pressed his bulky body against her slenderness.

"Buena suerte," she had whispered into his ear.

Now, with the panic hitting his chest moments after the lack of air, he realized that he was trapped in the embrace of the young woman.

165 feet (6 atmospheres), 2 liters of air in his lungs: Fingers of his gloves working madly over his chest, Robert unclipped the fasteners on his BC and wrestled his cumbersome tanks off his shoulders, as if fighting his way out of her arms, so the tanks' bulk wouldn't drastically slow him on the ascent. Lungs already straining for breath, he furiously kicked for the surface with a single, panic-infused thought: to get away from here, from her, and get up there.

132 feet (5 atmospheres): With the lessening water pressure, the air in his lungs expanded to 2.4 liters. Robert didn't notice. Fins flipping like some baitfish under a predator's pursuit, the tables of hunter and hunted having turned, he raced upward past the shower of tiny bubbles that danced toward the surface from his last breath. A diver's rule of thumb is not to ascend faster than those tiny bubbles, about 30 feet per minute, to allow the nitrogen to seep from the tissues. Worse, he violated another essential scuba-diving rule: to keep breathing regularly on the ascent. These abstract rules were

overridden by a single basic instinct: to hold on to his air until he reached the top.

99 feet (4 atmospheres): The air in his lungs expanded to 3 liters, slightly easing his need to breathe. It also gave him extra buoyancy for his ascent, as did the expanding air pores in his wet suit. He shot toward the surface.

Scuba divers in advanced training practice emergency ascents from 60 feet on a single breath, and submariners in training ascend inside dive towers—tall tubes filled with water, with recompression chambers near at hand—on a single breath. Ironically, the problem is not lack of air but too much air. A single breath of compressed air inhaled from a scuba tank at 60 feet will expand in the lungs with the lessening pressure to the equivalent of three breaths by the time the diver reaches the surface. The trick is to allow the expanding air to escape the lungs in a controlled manner while not exhaling the entire lungful at once, leaving nothing but seawater to breathe.

66 feet (3 atmospheres): The shimmering underside of the swells grew more distinct, their foaming crests visible as bubbly streaks that Robert had seen as chalky specks from the airplane at 33,000 feet. The surface suddenly looked possible for him to reach. The air in his lungs doubled from the original half breath to 4 liters, or nearly a full breath. But he had also kicked his way straight up through Haldane's ratio and the decompression stops he should have made to let the nitrogen exit his tissues. Instead, as he shot upward, he fizzed inside like a just-opened bottle of soda. Some bubbles formed in his joints, which, in a matter of minutes or hours after their formation, would cause the characteristic joint pain of the bends. Others formed in his skin—soon to cause the itching and rash known as the "skin bends." Other nitrogen bubbles forming in his tissues damaged his capillary walls, entered his veins, and grew in size as they were pumped through his circulatory system.

For reasons still unknown, the nitrogen bubbles also have an affinity for the white matter of the spine, which contains nerve fibers that carry messages between body and brain. The victim often feels a tingling or sense of constriction around the chest or abdomen upon surfacing. This is followed over the next thirty to ninety minutes by numbness, weakness, loss of bladder and bowel control, and sometimes pain in the back and chest. In severe cases of this neurologic decompression sickness, the victim sometimes loses consciousness and then awakens to discover that his or her legs are permanently paralyzed.

33 feet (2 atmospheres): The silver underside of the surface pulled Robert upward. The air in his lungs had expanded to 6 liters—as much as a very large breath—filling his lungs to maximum capacity. But this last 33 feet to the beckoning surface was in some ways the most dangerous underwater zone of all. Air doubles in volume as it expands from two atmospheres of pressure at 33 feet to one atmosphere at the surface. A diver holding his or her breath with full lungs needs only to ascend between 4 to 8 feet to damage them.

25 feet (1.8 atmospheres): There was no pain. If he hadn't been so panicked to reach the surface, Robert would have noticed the expanding fullness in his lungs. But he kept kicking hard, shooting up like an undersea missile with the buoyancy of the expanding air. In the last 25 feet to the surface, the air sacs in his lungs, called alveoli, ruptured without pain. By the time he reached the surface, the air in his lungs would have ballooned to 12 liters, double the size of his largest breath, if he could contain it. He couldn't. As he fought the last few feet to the top, a stream of bubbles forced its way up his air passages and out his mouth.

0 feet (sea level, 1 atmosphere): Robert broke the surface in a flurry of bubbles, his upper body shooting above the water before toppling back down like a surfacing whale. The wind and waves and bright tropical sunlight slapped suddenly against his face. It had been less than a minute since he left the bottom. He exhaled, gulped for air in a great gasp; as he did, he spotted Felix's boat about 300 feet away, rising and sinking on the swells. He managed to raise his arm to signal Felix to come pluck him out. Limp with exhaustion, Robert

slowly treaded water and tried to catch his breath. Inside his lungs, however, the broken capillaries of the ruptured alveoli were now exposed to the air he was sucking in with every fresh, eager breath. Besides the nitrogen bubbles already lodged in his tissues, air from his lungs entered the broken capillaries in small bubbles, pumped through the left side of his heart, then raced up toward his brain. The bubbles lodged in the tiny capillaries of his brain. They prevented the flow of blood, rich with oxygen borne in its hemoglobin, from delivering its vital load to his brain tissues. More and more air bubbles raced from his lungs toward his heart until it beat only on bloody foam. There was nothing but foam to pump up his aorta toward his brain. Within seconds of surfacing Robert lost consciousness.

By the time Felix sighted Robert, pulled the cord of the outboard motor, hauled up the anchor, and motored through the swells to the deep blue channel, Robert was no longer waving. He was facedown, rising and sinking with the motion of the swells, his wet suit keeping him afloat.

Felix, grunting and straining and with the help of the swells' lifting motion, managed to haul Robert's neoprene-clad bulk over the gunnel. Robert tumbled heavily into the boat's bilge like some great black fish.

Felix pulled off Robert's mask, peeled back his hood, and tugged down the zipper of the sleek black wet suit to expose his fleshy white chest, no longer moving. He scanned the surface for any gear that might have risen with Robert. Nothing. He checked again for vital signs. Nothing. The only thing of any interest that Felix knew of down in the deep cut of the reef was an old cargo barge that had struck the reef and sunk years earlier. Occasionally he or one of the other fishermen would inadvertently haul up a brick in one of the nets from its ballast scattered all across the bottom.

Felix then assessed the direction of the wind and swells, scanned the horizon for storm clouds, and checked the level of the gas tank, which sat in the bilge of the boat. The satchel of photocopied letters from the old woman and her beautiful granddaughter in Cádiz still lay in the bottom of the boat beside Robert's body. Felix flipped the gearshift into forward and swung the bow of the wooden boat south toward the island, running at half speed to avoid the pounding from the swells.

There was no reason to hurry.

CHAPTER TEN

IN LOVE'S

BLOOD:

CEREBRAL MALARIA

Their first night on the beach, Zach and Jason pitch their tent near some other travelers and walk beneath the coconut palms toward the only light. This is a little restaurant that an enterprising local family has set up to cater to the beach-loving foreigners. As they enter, dim orange bulbs under the thatched roof throb unsteadily to the beat of the local generator. A boom box puts out a steely, cascading music that sounds like a fusion of reggae and gamelan. Clumps of young Western travelers sit at benches around bamboo tables laden with bottles of Indonesian beer and plates of food, talking and laughing.

"Over there," says Zach quietly, gesturing with his chin toward an empty table.

Jason picks up the cue. Beside the empty table is one surrounded by three women, their hair sun- and salt-streaked, their skin tanned, their arms sinewy and muscular from stroking through the warm sea. After two weeks up in the highland jungle climbing volcanos, the closest thing approaching a social life that Jason and Zach have had has been a persistent band of monkeys hanging around their camp. Jason heads eagerly toward the table, thinking, *What luck!*

"Let's stay cool for a minute," whispers Zach as they sit down.

They've been traveling through the tropics for nearly six

months now, living on the cheap out of their backpacks in Fiji, Irian
Jaya, and now an island off the coast of Sumatra. They'd heard about
this spot all the way back in Fiji: a great beach, some good smoke, a
small colony of Western surfers and snorkelers and backpackers
looking for a laid-back time.

The teenage daughter of the proprietor comes around to their
table and takes their order of beer and fried noodles. As she speaks
her halting English, just behind her head, on a dried leaf of palm
that's woven into the underside of the eaves, sits a small brown
speck. Even if they were looking, Zach and Jason probably wouldn't
notice it, and they certainly wouldn't know that it is a female
anopheles mosquito—the carrier of the malaria parasite. Instead,
they are working hard to appear casual in front of the three women.
When the waitress returns with the beer, they each take a few quick
slugs from their bottles, and Zach casually leans over to the closest
woman.

"How's the snorkeling here?" he asks.

Fifteen days earlier, that small brown speck hatched from a
pupa lying atop a rainwater puddle near the village temple. Letting
her wings dry for a few hours in the daylight, she set out at dusk to
find a mate. Near the village meeting hall she heard the irresistible
buzzing wingbeats of a swarm of male anopheles. She and a male
threw themselves at each other and copulated, he injecting his sperm
into her followed by a secretion that seals her entrance to any other
male's sperm. She now carried all the sperm she needed to fertilize
her entire lifetime's egg production—a few weeks, a month, maybe
more. Separating, the spent male anopheles flew off in search of a
refreshing drink of fruit juices; the female, in contrast, took to the
air to sniff out a meal of human or animal blood proteins to nourish
her eggs.

"The snorkeling's great, mate," the woman at the next table is
telling Zach. "Come on out on the reef tomorrow and see for yourself."

"I'll do that," says Zach. "I'll be there."

She's Australian, it's plain from her accent, and seems very friendly. Her blue sarong matches her lively blue eyes. Her feet are bare in the powdery sand. Zach gives a little knowing look to Jason that conveys, *This situation could be good for both of us, but let me handle it to start.* Jason gives a little nod of acknowledgment.

Zach leans closer. His skin has begun to flush with beer and tropical heat and restrained eagerness. "What's your name?"

"Chloe," she replies.

The two other women are talking about the embroidered Sumatran fabrics that they saw that morning in the market town a 6-mile motorcycle ride away.

"Have you been here a long time?" Zach asks Chloe.

"Not long enough," she says. "Two months. I came here from Perth with my boyfriend, but he's so particular about his little domestic comforts that he couldn't stand living on the beach and went home. I said to him, 'Rack off, then,' and decided to stay."

What an idiot! Jason is thinking about the boyfriend.

Abandoning their trysting spot at the village temple, the female mosquito flew off to a nearby family compound and found a teenage boy curled up on a sleeping platform. From the exposed skin behind his knee she sucked up a stomachful of his blood. His blood happened to be carrying *Plasmodium falciparum*, the tiny parasite that is responsible for the deadliest strain of malaria. Like many dwellers in endemic malaria zones, he had been exposed to it repeatedly since childhood and, though other village children died of it at an early age, he, like other survivors, developed a partial resistance to it known as "premunition." The malaria manifested itself in the boy in intermittent fevers, weight loss, and an enlarged liver and spleen. If this village were in sub-Saharan Africa instead of Asia, however, some of the village children would be protected from *P. falciparum* malaria by an abnormality in their blood hemoglobin, known as the sickle-cell trait.

As the village boy's blood was sucked up into the mosquito's proboscis and slid down into her stomach, his red blood cells burst open to release *P. falciparum* parasites in their sexual form. Inside the mosquito's tiny stomach, the male *P. falciparum* sperm, their whipping tails propelling them relentlessly forward, drove at the female eggs, connected, then merged. These fertilized *P. falciparum* eggs, in one of the infinitesimally tiny but profound engineering marvels of the malarial world, shaped themselves into little augurs and drilled their way out of the mosquito's stomach. There they attached themselves to the gut's outer wall and swelled up like water balloons filled with hundreds of tiny eel-like creatures called sporozoites. The cysts burst, and thousands of *P. falciparum* sporozoites flooded out, swimming directly for the mosquito's salivary glands.

This all took at least two weeks from the time the mosquito first ingested the parasite in the village boy's blood. During those two weeks, she'd laid a batch of her own eggs, replenished herself with human blood, and laid more. But these earlier bites to humans were not infectious; the *P. falciparum* parasites, in their various forms, were contained in her stomach or encased in cysts on her stomach's outer wall. But now, as Jason and Zach sit at the table beneath her perch in the palm-thatch roof, the *P. falciparum* sporozoites have burst forth from their cysts and are swarming through her salivary tubes.

"Do you mind if I join you?" Zach says to Chloe, sliding toward her along the bench, while sliding along the table his tall bottle of Bintang beer.

"Please do," she says to Zach, and then nods toward Jason. "And tell your friend he's welcome to join us, too."

As Zach slides onto Chloe's bench, his billowing exhalations of carbon dioxide and the waves of body heat from both their elevated skin temperatures rise toward the palm-thatch eaves. The glowing body heat and clouds of carbon dioxide catch the attention of the small brown speck perched on the dried palm leaf. Instinct tells the anopheles female that human blood is rising close to the skin surface somewhere nearby. She craves the blood to nourish yet another batch of her eggs.

She springs from her perch and flitters down toward the table, unobtrusive and unseen in the dim light. She circles above their heads, first the two women, who emit the least heat and carbon dioxide, then to Jason, who she senses is warmer, and to Chloe, warmer still, and finally to Zach, wrapped in a rich cloud of carbon dioxide and emitting intense infrared rays from an exposed V-shaped patch of skin on the side of his neck outlined by two stiff ropes of his Rasta braids. She lands lightly on her six legs—too lightly for him to notice. Zach is animatedly telling Chloe about his adventures climbing the volcano with Jason, and how an explosive hail of lava rock nearly struck them down at the summit. He makes an explosive gesture with his hands to show the dire danger they were in from the lava bombs, and just then the anopheles mosquito thrusts her tiny proboscis into his neck.

Malaria has something of a romantic reputation, at least here in the temperate regions of the globe. Eradicated in the United States nearly a century ago, it remains in many of our minds the disease of explorers and adventurers, Amazonian missionaries and African ivory merchants. How many Hollywood jungle dramas have we seen where the hero is lying by lamplight in a thatched hut, delirious with fever and face shiny with sweat while his faithful retainers look on anxiously? After all, it was a jungle fever—no doubt malaria—that finally felled that most magnificent specimen of these adventurers, the ivory-trader-gone-native Kurtz in Joseph Conrad's *Heart of Darkness*.

The Congo River setting for Kurtz's malarial demise couldn't have been more appropriate, because it is believed that the malaria parasite *Plasmodium* (the name refers to a genus that contains several species) first adapted to human hosts during the Neolithic period in this very region—the same exotic-disease hothouse that much more recently spawned the AIDS and Ebola viruses—as humans encroached on jungle lands for farming. From there, migrating peoples unwittingly carried the parasite to Europe, Asia, and the Americas, where the anopheles mosquito already lived. Malaria is thought to have arrived in the Americas, somehow, several centuries before Columbus. During periods of human history, malaria has been a great scourge, not unlike the Black Plague or the AIDS epidemic, and its outbreaks have altered the course of history. Malarial fever decimated the cream of the Athenian army in 413 B.C. when its soldiers camped in a swamp while attempting to lay siege to Syracuse, and Athens never fully recovered its power or prominence. Likewise, it was probably cerebral malaria that killed the otherwise indomitable Alexander the Great in 323 B.C. at age thirty-three in the city of Babylon on the Euphrates River, and with his death the enormous empire he had assembled through conquest all the way to India quickly came unraveled. The great Italian poet Dante Alighieri died of malaria, Holy Roman Emperor Charles V succumbed to it in Spain in 1558, and the American colonial settlers in Jamestown struggled against several different strains of malaria brought by English ships and African slaves.

For centuries, malaria's actual cause and means of transmission remained a mystery, but humans always associated it with damp, low places. Shakespeare, whose contemporaries knew the periodic malarial fevers as "agues," has his Caliban wish on Prospero "all the infections that the sun sucks up / From bogs, fens, flats . . ." Surrounded by swamps, it was the citizens of Rome who gave the disease its current name—*mal aria*, or "bad air." Up until the end of the nineteenth century it spread far beyond the tropics, summer epidemics sweeping up into temperate regions as far north as Holland, to the Russian city of Archangel, just below the Arctic Circle, and throughout the more moist regions of the United States and southern Canada.

In 1880, Alphonse Lavaran, a French army doctor living in Algeria, spotted through his microscope tiny parasites in the blood of his malaria patients. Seventeen years later, Ronald Ross, a surgeon-major in the Indian army, discovered that the parasite was transmitted by a very specific type of mosquito—the small, unobtrusive members of the genus *Anopheles*. Western doctors had known since the early seventeenth century of a treatment for malaria—quinine, derived from the bark of Peru's cinchona tree, a secret known by Inca herbalists who conveyed it to Jesuit missionaries. But with the discovery of the mosquito transmission of malaria—for which Ronald Ross won the Nobel prize in 1902—hopes soared that it might be eradicated from the face of the earth.

Swamps were drained, insecticides sprayed, villagers educated, and new drugs invented that mimicked, with greater potency, quinine's effects. This battle wiped out malaria in almost all of Europe, the United States, and other temperate regions. But the disease not only lingered on stubbornly in the tropics, but continued to thrive and at times explode as the stubborn mosquito and clever parasite developed resistance to new insecticides and drugs. In India, 100 million people suffered from malaria in 1952; ten years later, with the spraying of DDT on mosquito-breeding areas, this plummeted to 60,000, only to rebound into the millions soon after. Today, some 3 billion of the world's people—roughly half—live in malarial areas. The infection rate among children in some parts of Africa exceeds 50 percent, many of whom will die. The World Health Organization estimates that 300 million cases of malaria appear each year, and 1.1 million people die of it. In terms of mortality, malaria reigns as the world's third-leading infectious and parasitic disease behind tuberculosis (1.6 million deaths annually) and AIDS (2.7 million deaths).

To temperate-region dwellers, malaria may be a disease tinged with the romance of exotic jungles but to those who live in the tropics it is an everyday fact of life. Most strains of malaria cause a form of fever and illness no more dangerous, at least according to some authorities, than a case of influenza. But one strain of the disease is deadly among the native populations, especially in small children ages one to five who have not yet developed a tolerance to it, as well as in travelers from the temperate regions. This strain of the malaria parasite, called *Plasmodium falciparum*, can be treated easily in its early stages, though its symptoms are difficult to read and often mimic other diseases. If they are prudent and know of the danger, travelers can take preventive antimalarial drugs before entering the tropics. Left untreated, it can develop into severe or cerebral malaria, however, and it fully merits the mysterious last utterance of Conrad's Kurtz. Instead of referring to the fallen state of his soul or the unmitigated rapacity of European civilization in the Congo, his words "The horror! The horror!" could just as easily have been describing the rampages of *Plasmodium falciparum* throughout his bloodstream and brain.

As Zach talks animatedly to Chloe in the palm-thatch restaurant, special cutting tools on the proboscis that are designed for human skin slice down through Zach's epidermis, into his underlying dermis layer and the thin layer of subcutaneous fat beneath. Here, a network of tiny capillaries branch out like the limbs of a tree toward the skin surface. They have dilated with Zach's flushing and are pumped full by the excited beat of his heart. The proboscis penetrates a capillary and strikes Zach's blood. The mosquito contracts her glands to squirt saliva into his blood to prevent it from coagulating as she sucks it into her gut. With her saliva, scores of the tiny *P. falciparum* sporozoites swim into Zach's blood vessels.

"We thought we were dead." Zach is telling Chloe about their volcano-climbing adventure. "Lava bombs dropping everywhere—*whump, whump, whump!*"

He makes splattering gestures with his hands on the table. Chloe laughs appreciatively.

"We ran for our lives," he says. "We started at ten thousand feet on the summit and we didn't stop running until we were down to five."

The women laugh at the image of Zach and Jason sprinting downhill through the jungle-covered mountain flanks with lava bombs chasing them. Had they stayed up near the summit, however, and taken their chances with the lava bombs, there would be no sporozoites now swimming in Zach's blood vessels. Malaria's delicate and complicated transmission cycle requires a steady, warm temperature averaging at least 60 degrees for a solid month to keep the mosquito alive long enough for the parasite to grow fully within her. Even on the equator, malaria can't be transmitted above 10,000 feet because it's too cold there.

Zach hardly feels the bite. He unconsciously scratches at the side of his neck between his Rasta braids. By then, the anopheles mosquito, heavily laden with his blood in her gut, has hefted into the air and returned to her perch on the underside of the thatched eaves.

"Hey, I have an idea," says Zach. He's feeling itchy and hot in the tropical night. "Let's all go for a swim in the lagoon."

"Let's do!" says Chloe. "The phosphorescence is phenomenal!"

The others assent, and they pay their bill and spill forth from the thatched restaurant into the warm, buzzing night, leaving the female anopheles slowly and contentedly digesting Zach's proteins under the eaves. Here she will remain for three days, shaded from the sun and close to her feasting site, as she digests Zach's blood proteins and her eggs grow. Then she will fly off to lay them on a nearby pool of fresh water to complete her reproductive cycle.

The swimmers strip off their clothes in the dark and pile them on the sand. Zach is thinking that this night he might have the chance to complete a reproductive act of his own. They plunge into the sandy lagoon in shallow dives, watching each other underwater as the phosphorescence streaking off their bubbling bodies leaves glowing streaks like comets. Fish scatter, leaving their own quick flicks of phosphorescence as swirling fins agitate the light-emitting microorganisms.

As they frolic in the dark lagoon, diving under each other, over each other, between each other's legs—just another clump of sexually charged organisms in the wide, fecund, life pool of the tropical sea—the *P. falciparum* parasites are doing the same in Zach's blood, with even greater single-mindedness and without the immediate need for a partner. They'll carry their traveling party to various destinations in Zach's body over the next nine days. The first hot spot to which they head is his liver, which they need to reach immediately or they perish, and they spend some time there to let the party gather steam. Upon arrival, each sporozoite drills its way into one of Zach's liver cells. Cloaked within the cell walls, the sporozoite changes form and undergoes an orgy of reproduction, dividing again and again—the party suddenly gone mad—until finally the liver cells burst open. Each ruptured cell releases 10,000 to 30,000 tiny swarming creatures with the frightening name of invasive merozoites.

The hordes of merozoites now spill out of Zach's liver into his bloodstream. They are marching to a mission: to colonize as many of Zach's red blood cells as possible so that if some hungry female anopheles mosquito should again land on him for a blood meal, she'll be sure to suck up enough of Zach's parasitized blood cells to kick off *P. falciparum*'s reproduction cycle in her stomach and spread it to another human. This is how *P. falciparum* keeps itself alive as a species, although sometimes in its zealousness to survive it kills the human host that feeds it.

Once they've escaped from his liver and are afloat in his blood, the pear-shaped invasive merozoites work with bluntness and efficiency, shoving with their conical noses through the walls of Zach's red blood cells in the remarkably short time of twenty to thirty seconds after initial contact. Shedding their outer cloak roughly upon entry like an uninvited guest, they change form, feast on a groaning banquet of cytoplasm and hemoglobin, then divide into eight to sixteen new merozoites that crowd the blood cell to the bursting point and heartily earn the name parasites. Finally the red blood cell, or what's left of it after its been eaten away from inside, bursts open and releases this new family of invasive merozoites. They each invade other red blood cells, produce more offspring, and the cycle continues, as the numbers of invaded cells and *P. falciparum* parasites in Zach's blood rise astronomically.

Jason hardly sees Zach during the next nine days. The midnight skinny-dipping party breaks up when Zach and Chloe climb dripping and naked out of the water, gather up their clothes, and walk alone together down the beach. The frolicking spirit suddenly drained from the party by this private liaison, Jason and the two other women dress awkwardly and disperse to their separate tents, Jason all the while recriminating himself for not being more aggressive like Zach. Another missed chance, he thinks. He always misses his chance. Zach is always the chosen one.

Zach doesn't return to their tent that night, nor the next nor the next. By day, Jason sees him with Chloe on the beach or snorkeling on the reef, while he's left glumly to entertain himself. The invasive merozoites, meanwhile, are furiously busy beating the streets of Zach's circulatory system, taking it over cell by cell. On the ninth day, Jason is lying under the shade of a palm tree reading—for the second time—a collection of Somerset Maugham stories about British colonists washed up in the tropics that he purchased at a bookstore back in Singapore. He notices Zach's shadow fall across the page.

"So where's Chloe?" Jason says without looking up.

"She left," Zach replies.

Jason puts down the book in the sand. "She *left*? You're joking, aren't you? Why did she leave? I thought she wanted to stay here forever. Especially since she met *you*."

Zach flops down on the sand beside him, lies on his back with his Rasta braids splayed, and stares up into the palms. His fingers absently work the grains of sand clinging to his browned chest.

"She said she misses her boyfriend back in Australia," he says absently.

"Bummer, man," says Jason. "After all this, she wants to go back to her boyfriend? I was getting ready to pack up and go on without you. Well, I'm glad that's over. At least now maybe we can have some fun."

"Yeah, whatever," replies Zach, closing his eyes. "But, man, I'm *really* tired."

In the night, Zach becomes chilled in his green nylon bag. Jason, holding the flashlight in his mouth, pries a small yellow pill from its cardboard casing and hands it to Zach lying beside him. Zach's hand emerges from the green cocoon. He tries to grasp the pill, but his hand trembles violently and he drops it. Jason picks it up, drops it on Zach's tongue, gives him a sip from his water bottle.

The pills, antihistamines, have worked every other time when they couldn't sleep, tossing uncomfortably in the tropical nights with insect bites or sunburn or coral cuts on their feet. Instead of antihistamines, however, what they should have been taking was a preventive dose, weekly or daily depending on the type, of one of the antimalarial drugs.

Zach curls up on his side, hugging himself with his arms inside his bag to stay warm. Paradoxically, this feeling of chill and shivers is the body's response to a rapidly rising internal temperature—the onset of fever—which will play a therapeutic role: to try to cook the invading parasites to death.

When confronted by bacteria, viruses, or other foreign invaders, the white blood cells in the human circulatory system release proteins, known as pyrogens, which cause a series of chemical reactions that trigger the body's thermostat, located in the hypothalamus portion of the brain, to dial its temperature up to a higher setting. The malaria parasite floating about in the blood is so effective at inducing these fevers that in the 1920s, before the advent of penicillin, blood infected with *Plasmodium falciparum* parasites was injected straight into the frontal lobes of syphilis patients. The resulting "therapeutic malaria" fever proved effective in killing off the syphilis bacteria in 80 percent of the cases, though the malaria parasite itself survived the fevers intact and had to be treated separately.

Jason switches off the flashlight and lies back on top of his bag. As he falls asleep, he can hear the rustle of nylon as Zach twists and curls about in his bag with the chills and shivers. These help generate muscular heat to elevate his temperature to the new setting.

With each 1-degree rise in his temperature above 98.6, Zach's metabolic rate—his respiration rate, heart rate, blood flow—jumps by 7 percent.

101 degrees: mild fever . . .

102 degrees: the body's metabolism is hopped up some 30 percent higher than its normal rate, causing extreme nervousness and irritability . . .

An hour after falling asleep, Jason is woken by a thrashing of arms and legs in the sleeping bag beside him.

"Give me some air," Zach cries out. "I'm really hot."

Jason, sitting up in alarm, unzips the flap of the tent.

"Do you want to go for a swim?" he asks.

"Not a swim," says Zach, recoiling. "Fuck the swim!" The idea of the seawater on his burning skin makes him think that he might explode like a hot rock tossed into cold water. His head is pounding with pain and his neck is stiff. He twists about on top of his bag, trying to find a cooler spot, and finally drags himself on elbows and knees out onto the cool, powdery sand like a lizard regulating its body temperature by lying in sun or shade.

103 degrees: confusion . . .

104 degrees: delirium . . .

Zach's internal temperature hits 105.2 degrees Fahrenheit, enough to kill syphilis and many other bacteria by starving them of the blood iron they need, but not the malaria parasite itself. Jason makes a bed for him under the stars, then drags out his own bag and falls asleep beside him. Zach neither wakes nor sleeps but tosses in the netherworld of fever dreams, not knowing where he is, only that he's beneath the night sky and the stars, an overheated lizard writhing on the earth, tail lashing about, until, toward dawn, sweat suddenly pours out over his skin, bathing him, cooling him. Zach dimly begins to remember that he is on an island off the coast of Sumatra and that the lump beside him isn't a rock or another lizard but his friend Jason. He then falls into a deep, exhausted sleep.

Zach's first fever bout has run the typical course noted by physicians since ancient times: chills, a quickly soaring temperature, a feeling of crippling heat for one to two hours, and then a pouring sweat that indicates a falling temperature and is so familiar from the tense bedside vigils of the Hollywood jungle dramas—"Tell the natives! Bwana's fever has broken!"—until the body's temperature drops back to normal five to eight hours after the attack began and the victim lapses into exhaustion.

They awaken late, with the sun filtering through the palm leaves onto the white sand. Zach's fever has totally disappeared. He feels fine, although tired.

Jason sits up, rubbing his eyes against the sunlight. "Maybe it's something you ate," he says helpfully.

"Yeah," says Zach. "Tonight, remind me to stay away from those Sumatran chili peppers."

They both laugh. It'll make another good story when they get home.

That day they spend quietly, with Zach resting near the tent and reading Jason's copy of the Somerset Maugham stories. They'll go to the caves tomorrow, they decide. Zach's fever hasn't really ended at all, however; it's only that the *Plasmodium falciparum* parasites in his red blood cells have taken a short breather. The *P. falciparum* parasite splits on a forty-eight-hour cycle, from the time a merozoite invades a red blood cell to the time the eight or sixteen new merozoites burst from the cell to invade new cells. Because the merozoites manage this in coordinated fashion, Zach's fever thus will attack him at forty-eight-hour intervals—a bout of fever on the first day, a break on the second day, fever on the third day, a break on the fourth, fever on the fifth day, and so on. This every-other-day fever cycle is what's known as tertian malaria. It is characteristic of three of the four strains of the malaria parasite that infect humans, though the clocklike cycles often blur in the case of *P. falciparum*, and in the case of *P. vivax* they can appear years after the initial infection.

That night Zach sleeps soundly beside Jason in the tent and experiences no fever. The next morning he feels well enough. He is still weak, however, because so many of his red blood cells have burst that he is becoming anemic. Trembling slightly, he walks with Jason along a sandy path into the village, and they find two young men with motorcycles willing, for a price, to transport them into the hills and guide them to the caves. For much of the day, trailing the village youths, Jason and Zach slip and slide along muddy paths. At the base of white limestone cliffs overhung with curtains of vines, dark apertures suddenly appear in the earth. With their camping headlamps and a couple of candles, the foursome poke a short distance into the caves, marveling at the stalactites hanging from the ceiling that glisten in the light of their headlamps and the thousands of inverted bats that cling to the cave roof, their wings folded in sleep, awaiting the wakening of the tropical night.

In similar fashion, Zach's parasitized red blood cells have begun to adhere to the ceilings of the tiny capillaries in his brain. This is the beginning of the highly dangerous condition known as cerebral malaria. The *P. falciparum* parasite is unique in that the red blood cells invaded by the parasites develop tiny knobs on their surface containing a protein that causes them to stick to the cells of the blood vessels. They have a particular tendency to migrate to the blood vessels in the brain and stick there. Of the many bizarre properties of malaria, one is that this process of sequestration appears suppressed among inhabitants of malarial regions who have developed a tolerance to the disease, and—no one knows quite why this is so—also among individuals who are missing their spleens.

Likewise, the sickle-cell trait fights sequestration because the abnormal hemoglobin inside the red blood cell that is invaded by parasites bundles up into a stiff mass and destroys the cell. Thus the sickle-cell trait is thought by geneticists to be a preventive adaptation to malaria among sub-Saharan Africans, where the deadly *P. falciparum* form accounts for the majority of the disease.

But Zach is from the temperate north; he possesses a spleen and normal hemoglobin, and he hasn't been taking malaria pills. Toward late afternoon he starts shivering again. At first he thinks it's only the cave's cold and damp, but the chills and shakes don't stop when he crawls out into the muggy, leafy daylight. "Fuck!" he whispers to himself, his lips trembling, as he sits on the ground and hugs his knees to his chest. After a shaky ride back to the beach, Jason and the village youths help him from the motorcycle saddle and lower him gently onto the sand near the tent. Once again, the chills give way to the unbearable heat, until midnight, when Zach's tossing fever finally breaks into sweat, and then he sleeps. Jason, keeping a vigil, is truly alarmed; this is something far more serious than an overdose of Sumatran chili peppers.

After his own restless night, Jason is already awake and squatting on the sand near their camp stove brewing tea when he notices Zach open his eyes the next morning.

"What do you want to do?" Jason says quietly. He stirs sugar into the pot, trying to hide his alarm.

"I don't know," says Zach without raising his head, staring out toward the breakers crashing on the reef. "I don't know how many more nights like that I can take."

"Where do you want to go?" asks Jason, trying to sound neutral. He wants Zach to come to the decision that he already has.

"Wherever there's a decent hospital, I guess. Back to Singapore."

"Good," says Jason, turning off the little knob of the stove. "Let's do it."

The ferry takes twelve hours back to the main island of Sumatra, then they must ride six hours by bus to the opposite shore and board another ferry to Malaysia, followed by a train south to Singapore. By now Zach's fever is not arriving at forty-eight-hour intervals but seems to rise and fall in wild, erratic swings. He's fainted twice upon standing, once getting off the bus and again getting on the train, because the blood supply to his brain is slowly being choked off by knobby red blood cells packed with *P. falciparum* parasites sticking to the walls of the capillaries. As his brain loses oxygen, odd personality changes occur. He becomes inexplicably irritable toward Jason, who's hauling both their backpacks and propping up Zach at the same time.

"Take it easy, dude," Jason says gently. "I'm only trying to help you."

On the final train run into Singapore, Zach's head lolls against the back of the compartment seat, eyes half open and unseeing, and his arms and legs twitch and spasm sporadically. Jason must hire a porter to help carry Zach from the train. He looks like a drunk, staggering half conscious under their support through the station; in regions where it is not common, cerebral malaria is sometimes mistaken for other afflictions such as epilepsy and meningitis, or for alcohol intoxication. But here they know about malaria. When their taxi reaches the hospital, Zach is totally unconscious. He's rushed into the emergency room, and the nurses and orderlies and doctors who examine him and question Jason immediately suspect the cause. They draw blood, it's dispatched to the lab, and the slide is quickly examined by microscope. The iron from the hemoglobin digested by the *P. falciparum* parasite has left malaria's characteristic dark pigment.

A short time later, a doctor hurries into the waiting room. She's a tall middle-aged German woman who speaks English with a thick accent.

"Why have you waited so long to bring him here?" she barks out at Jason. "His blood is *black* with malaria!"

She then turns and rushes back.

Zach lies in the intensive care unit. An oxygen mask is over his face to increase the oxygen flow to his choked brain. The doctor has ordered an IV in his arm and a large loading dose of quinine pumped into his circulatory system over a thirty-minute span, to be followed by a maintenance dose at eight-hour intervals. The doctors could have chosen from other, more modern, synthetic antimalarial drugs, but quinine still remains one of the most effective, nearly four centuries after missionaries learned of it from the Incas and many decades since the British in India discovered they could disguise its bitter taste by mixing it with gin, thus giving birth to the gin and tonic. The first widely used modern antimalarial drug, chloroquine, was discovered by German scientists in the 1930s and helped protect Allied forces in Asia during World War II. During that war and in the years that followed, in an enormous battle to combat malaria, some half million substances were tested as antimalarial drugs, but only a few were found to work. In some regions the always inventive *P. falciparum* parasite developed a resistance to chloroquine. In these areas, doctors must rely on newer drugs such as mefloquine or fall back on the old standby, quinine. Derivatives from a traditional Chinese medicinal herb—*qing hao*—have also been shown in clinical trials to clear the parasites rapidly.

As he breathes slowly through the oxygen mask, thick wads of parasitized red blood cells clog in the tiny capillaries of Zach's brain like fallen leaves plugging a rainspout. The quinine coursing into his circulatory system from the IV attacks the parasite's nucleus and cytoplasm, but the *P. falciparum* parasite now infects over 30 percent of his total red blood cells.

Zach is in a deep coma. Jason sits in the waiting room, trying to read his Somerset Maugham stories. Across the room, Singaporean children tumble about their parents' legs. Jason can't concentrate on the story, wondering if he could have done more to help. How stupid of them, he thinks. Malaria. Why didn't they think of that days ago? Or would days ago have mattered?

The German doctor now pushes slowly out the door to the waiting room. She sits down beside Jason, and Jason already knows what she is going to say.

CHAPTER ELEVEN

IN A LAND

BEYOND

THE SHADE:

DEHYDRATION

You don't know why they're doing this to you. You don't even know exactly what it is that they've done. All you know—here in the middle of the Sahara Desert—is this:

In early evening the caravan stopped. They dismounted from their camels and walked barefoot in their flowing blue robes to your mount. You couldn't guess what they were thinking; their traditional turbans and blue veils—*tagilmusts*—wrapped across their faces revealed only keen, dark eyes. They forced your camel to sit. Silently, they removed your wristwatch and with rough, hand-plaited grass rope lashed your wrists to the tall pommel, then knotted a *tagilmust* as a blindfold around your eyes. You could hear the mutters and protests of the camels as they remounted. Then, taking your camel by its halter, they led you deep into the desert night.

You guess it's now past midnight. Other nights when the caravan has traveled so late, you felt this same cooling of the Saharan air and the aching soreness of your back from the hours of rocking on the camel's back. It's been two weeks since you've joined their caravan, five days since you left the last oasis with them, when they summoned you from the interview you were conducting under the palms with the marabout—the holy man—and told you the caravan was moving again. They're supposed to be taking you to the nomad guerrilla fighters out in the desert who have been conducting the raids on government outposts. Why are they doing this to you? And why now? Are they trying to scare you into keeping their secrets? Or prevent you from seeing how to get to the guerrilla strongholds? Or is it that they think you already know too much and plan to kill you? They know you're sympathetic to their struggle against a government far more powerful than they, one that wants to settle them, rule them, steal their freedom. That's the name they've always called themselves as a people—Imazighen, which means "free." To their sedentary neighbors at the Sahara's edge, however, they are known by another name, one meaning "those who are cast out by God."

A fashionable New York magazine has dispatched you here. The editor, leaning conspiratorially over the table during lunch at a trendy downtown trattoria, told you that the assignment could make your career as a freelance writer.

"And let's face it," he'd said, forking up with a deft spin another ample bite of spaghettini with black truffle sauce and washing it down with a sip of Barolo wine, "you've been struggling. Your career could use the help."

You couldn't argue with that. You *had* been struggling. The magazines no longer wanted to buy your lengthy, insightful articles about hard travel to remote places. They all clamored for upscale travel pieces—preferably involving Mediterranean countries and celebrity sunbathers (especially those sans tops) on large yachts, plus plentiful shopping opportunities. But when you'd called up the editor and proposed writing a piece about the nomads' guerrilla war in the southern Sahara Desert—a place you'd always wanted to experience—it caught his attention. He invited you to lunch, and you pitched it to him and he thought it over as he chewed his spaghettini and swilled his Barolo. Finally he told you the piece could forgo the shopping opportunities and celebrities as long as it promised the threat of violence. That's when he'd leaned conspiratorially over the table and said the piece not only would add a bit of *faux dangereux* thrill to the magazine's well-heeled image but could truly *make* your career.

And so in late April you flew to Paris and then boarded another flight full of tall men in elegant embroidered robes down to Ouagadougou, near desert's edge. It took over a week riding in hot buses and on the backs of trucks to get to the dusty provincial town where you first made clandestine contact with a representative of the nomad guerrillas. You explained what the guerrillas stood to gain by taking you into their confidence and out into the desert where they operated. Maybe your piece in the fashionable magazine would sway public opinion back at home. Maybe the nomad guerrillas would get help—foreign aid, perhaps even weapons, the small, potent, sophisticated handheld missile launchers that could knock a helicopter or even a fighter jet right out of the sky. The guerrilla nodded thoughtfully. He seemed to understand. One night he took you to the edge of town, where the mud walls gave way to sand, and introduced you to the nomads in the blue veils who were taking a caravan out into the desert. There were eight nomads, each mounted on his own camel and each leading a string of two or three other camels that carried loads draped over each side. The loads were small and heavy and wrapped in woven-straw matting. You didn't know what was inside them. You didn't want to ask.

"Go with them," he said. "They will take you to the people you wish to meet."

You immediately liked the nomads—their competence in the desert, their grace, their quick humor—and you thought they liked you. You could communicate with them, barely, in your broken French and the few words of their native Tamachek that you'd picked up. It had been a long, hard two weeks of travel in the desert, broken twice by stops at palmy, cool oases. All along, both in the oases and in the caravan, the people you'd met had showed you hospitality, and you sat with them late into the starry desert nights and drunk small glasses of sweetened green tea. You laughed and joked and tried to learn their songs. You shared their food. You passed goatskins full of lunch back and forth as you walked across the sands so as not to let the camels stop and lie down, each of you taking long gulps of the *aragira*—a slurry of pounded millet, dry goat cheese, dates, and water. But even then they never showed you their faces, drinking their tea or *aragira* from underneath the scarflike wrap. It is the men, not the women, who wear veils in this nomad culture; it is considered impolite for a man even to show so much as his nose. Instead there are their gaits and their eyes to distinguish them—eyes that are sometimes laughing, sometimes angry, always bright and acute.

"What are you doing to me?" you now call out from behind your blindfold.

But in reply you hear only the creak of their heavy loads as the camels plod slowly onward over the night-cooled sands.

They stop only once, to open a *gerba*—a goatskin bag that looks like a large football with a leather spout—dangling from a camel. You hear the sloshing of the water as it pours into a wooden bowl. They offer the bowl to you first, holding it to your lips.

"Drink," one says.

Swallowing gulp after gulp, you drain the entire contents of the bowl to slake the desert thirst that never leaves, only lessens. The water is wonderfully cool and refreshing despite the goat hair from the *gerba* and flecks of camel shit from the pool of water back at the oasis. Drinking the cool, refreshing water from the wooden bowl held to your lips is the last thing you remember.

You don't know where you are. You're lying on your side. First you see a rose-colored sky. Then, at eye level, an expanse of dull orange sand whose grains are as fine and unadulterated by pebbles, rocks, or anything else as a vast pile of table salt. You begin to remember: You're in the desert. You lie still. You listen for familiar sounds—the murmur of sleepy voices, the clank of a cooking pot, the mutter of a camel. Nothing. Not a chirp of a bird. Not a whisper of wind. Not a drip of water. There is only silence. Utter, absolute, stone-dry silence.

Alarmed, you sit up. Someone has draped a blue robe over you to serve as a blanket and placed beneath your head a mounded *tagilmust* for a pillow. But where are the people who brought you here? Surely they are nearby. You look around. Nothing but the fine orange sand cresting in low dunes. Overhead, the scraggly branches of a single acacia tree under which you've been placed. Then you notice the camel tracks. They lead off in seven or eight different directions where the nomads of the caravan apparently split up, each heading his own way purposefully over the naked dunes, as if to confuse you should you try to follow them.

It's then that you have a single, all-powerful thought: *Where's the* gerba?

At dawn on August 23, 1905, a bellow echoing up the canyon awakened naturalist W. J. McGee from his camp alongside some pools in a gorge in Arizona's remote Gila Desert, where he'd been studying the effects of light on desert life. McGee and the Papago Indian he'd hired as a tracker, Jose, sprang from their bedrolls to investigate. A quarter mile down the trail, they came upon Pablo Valencia—a formerly robust forty-year-old, 155-pound prospector who had passed through their camp eight days earlier in search of a "lost mine." Or rather, they came upon what McGee later described as "the wreck of Pablo."

"Pablo," wrote McGee with the precise observation of the naturalist in what has become a classic account in the literature of thirst, "was stark naked; his formerly full-muscled legs and arms were shrunken and scrawny; his ribs ridged out like those of a starveling horse; his habitually plethoric abdomen was drawn in almost against his vertebral column; his lips had disappeared as if amputated, leaving low edges of blackened tissue; his teeth and gums projected like those of a skinned animal, but the flesh was black and dry as a hank of jerky; his nose was withered and shrunken to half its length, the nostril-lining showing black; his eyes were set in a winkless stare, with surrounding skin so contracted as to expose the conjunctiva, itself black as the gums; his face was dark . . . and his skin generally turned a ghastly purplish yet ashen gray, with great livid blotches and streaks; his lower legs and feet, with forearms and hands, were torn and scratched by contact with thorns and sharp rocks, yet even the freshest cuts were as so many scratches in dry leather, without trace of blood or serum; his joints and bones stood out like those of a wasted sickling, though the skin clung to them in a way suggesting shrunken rawhide used in repairing a broken wheel. From inspection and handling, I estimated his weight at 115 to 120 pounds. We soon found him deaf to all but loud sounds, and so blind as to distinguish nothing save light and dark. The mucus membrane lining mouth and throat was shriveled, cracked, and blackened, and his tongue shrunken to a mere bunch of black integument. His respiration was slow, spasmodic, and accompanied by a deep guttural moaning or roaring—the sound that had awakened us a quarter of a mile away. His extremities were cold as the surrounding air; no pulsation could be detected at the wrists, and there was apparently little if any circulation beyond the knees and elbows; the heartbeat was slow, irregular, fluttering, and almost ceasing in the longer intervals between the stertorous breathings."[1]

Although he was at first unable to speak or even swallow, Pablo was slowly brought back to life over several days by the ministering of McGee and Jose, improving immensely several days later in Yuma after an all-day watermelon-eating and sleeping binge. Like many

---

1. W. J. McGee, as quoted in *Thirst: Physiology of the Urge to Drink and Problems of Water Lack* by A. V. Wolf (Springfield, Illinois: Charles C. Thomas Publisher, 1958).

victims of what's known as "desert thirst," Pablo had thrown away
every possession he carried during his waterless 100- to 150-mile or-
deal, the result of a missed rendezvous with a partner who was supposed to bring full canteens. First he'd tossed away his gold nuggets,
followed by his food, his coat, then his trousers, whose pockets contained his money, his tobacco, and his knife—even though it was the
thought of stabbing his partner that helped propel him onward.
Then he left behind his hat until he carried only a canteen in which
to collect and drink his dwindling drops of urine. On the morning
before McGee and Jose found him, Pablo had knelt down in the
shade of a bush as the sun rose, said a final prayer, laid down facing
east, and, regretting his lack of consecrated water, made the sign of
the cross. Then, as far as he knew, he died. His body lay on the baking desert floor while some innermost part of him hovered nearby
watching the buzzards inspect his corpse. That night, however, the
desert mercifully cooled to the low 80s. Something stirred in Pablo.
Naked, unable to see but feeling with his hands along the old settlers' trail near the Mexican border known as El Camino del Diablo—
"Road of the Devil"—he struggled through the night toward McGee's
camp and water, where he collapsed in the dust at dawn.

There is something mystical about the desert itself and the profound thirst that can so easily claim one there. Desert thirst is a hal-
lucinatory experience. Not only does one encounter the stereotypical
mirages—the shimmering "lakes" of water caused by sunlight reflecting off layers of air of different temperatures near the desert surface—but the victim deep in its throes sees visions, hears things,
undergoes out-of-body experiences. Like Pablo, the victim often
sheds all clothing and belongings and—having stripped oneself of
all that one has acquired on this earth—crawls naked through the
naked desert beneath the naked sky. Surely it is no coincidence that
the desert has long been visited by mystics, prophets, and seekers. Its
nakedness distilled and shaped some of the world's great religions.
Moses and the Israelites traversed desert realms in their forty years
wandering in the wilderness. Before he founded his own faith, Jesus

probably was a follower of the desert ascetic and wanderer John the Baptist. Muhammad's parents sent him as an infant away from the unhealthy climate of Mecca to live with a nomadic desert tribe. It is as if the desert strips away all superfluous layers—not only the rich, green layer of life that blankets so much of the planet, but also the superfluous layers of humans, of material acquisitions, of personality, even of ego itself.

The French, those great colonizers of the Sahara, have a term for the way in which the desert transforms a person: *baptême de la solitude*—"baptism of solitude." "It is a unique sensation," writes the novelist, musician, and Saharan wanderer Paul Bowles about standing alone in the desert night, "and it has nothing to do with loneliness, for loneliness presupposes memory. Here, in this wholly mineral landscape lighted by stars like flares, even memory disappears; nothing is left but your own breathing and the sound of your heart beating. A strange, and by no means pleasant, process of reintegration begins inside you, and [you] have the choice of fighting against it, and insisting on remaining the person you have always been, or letting it take its course. For no one who has stayed in the Sahara for a while is quite the same as when he came."

But the desert not only can profoundly transform the human spirit; it can also profoundly transform the human physiology. Humans are, by weight, nearly two-thirds water. Not only does the human body serve as an aquarium that bathes its living cells, as one French aquatic philosopher mused, but each cell itself is a tiny aquarium. One of the many remarkable properties of water is that it is a highly effective solvent—it easily dissolves many other substances. Thus in the form of aqueous bodily fluids such as blood plasma, which in humans is made up of over 90 percent water and transports nutrients, wastes, and heat through the body, water plays a role in almost every biological process of both plants and animals.

The aquarium or reservoir that is the human body is equipped with a precise system of valves to maintain its water level with extreme accuracy—within about .22 percent of one's body weight over a twenty-four-hour period, which in a 155-pound male translates to about half a cup. If too much water is taken in the reservoir's inlet—the mouth—the body's outlet valve, the kidney, will excrete the excess in the form of dilute urine. If too little water enters the body through the mouth or too much water dissipates from the reservoir—in the form of evaporating sweat, for instance—the level drops and the reservoir will close its outlet valve, allowing little urine to run out. But if the reservoir level continues to drop due to continued evaporation and lack of inflow—as can so easily happen in the desert heat and aridity—the reservoir's most distant, shallowest, and least important bays and arms start to dry up. Soon those distant arms and bays begin to expose their muddy, cracked bottoms. If the reservoir continues to shrink, all that remains is a muddy pool of sludge at its center that cannot support life.

Dipsologists—those who study thirst—measure a human's lack of water by calculating the loss as a percentage of body weight. An adult male weighing 155 pounds who is 60 percent water contains about 42 liters' worth. Under the most gentle of circumstances, he loses a minimum of about 1.5 liters of water each day—or about the contents of five big tumblers. Most disappears through urination, while about a third to half dissipates via sweat or the humid air from one's exhalations. Hot climates and hard work can boost water losses enormously to 1.5 liters or more *per hour*, mostly in the form of sweat to cool the body. A loss of 1.5 liters—whether it occurs over the course of an entire day or merely one very sweaty hour—represents about 2 percent of that male's body weight. The body can tolerate with only moderate problems a deficit of 3 to 4 percent of body weight, though one may suffer a powerful thirst, which kicks in at about a 0.8 percent deficit, or about a pint. At a 5 to 8 percent deficit—down about 3 to 5 liters—the victim becomes exhausted, complaining, prone to collapse, and at a 10 percent deficit—a very serious loss—physical and mental deterioration begins. By 12 percent—down about 8.5 liters—the victim can no longer swallow and is likely going into shock. Death occurs somewhere in the range of a 15 to 25 percent deficit. Or, put another way, death, in a previously healthy male weighing 155 pounds, occurs at a loss of 10 to 17 liters. Walking in the desert sun and sweating hard at 1.5 liters per hour, that moment could arrive in as little as seven hours.

"How long could I go without water, if by any chance I were left without it?" wrote J. S. Chase, a veteran traveler of the California deserts in the early twentieth century. "In that fierce heat, and struggling with that terrible country, a few minutes was as long as one could go without drinking. In a flash I saw what would be my condition *in a single hour*—torture: two hours—delirium: after that raving madness, till agony passed into insensibility, and that into death."

The classic studies of thirst conducted by E. F. Adolph and other U.S. military researchers in the deserts of southern California during World War II concluded that a person walking in 80-degree heat can travel 45 miles without water. A person walking in 100-degree heat can travel 15 miles without water. And a person walking in 120-degree heat—a temperature that is entirely possible in the Sahara and other deserts—will collapse after a mere 7 miles. Adolph's studies recommended as a general rule that the traveler on foot in the desert carry 1 gallon, or about 4 liters, for every 20 miles walked at night. To cover 20 miles on foot during the heat of the day, the desert traveler should carry 2 gallons. By this rule of thumb, to walk by day for 100 miles through the desert—an inconsequential distance in the vast Sahara, which, at 3.3 million square miles, could just about contain the entire United States—a traveler would have to haul on his or her back a staggering load of about 80 pounds of water. It is for precisely these reasons that it wasn't until the camel—a desert animal capable of walking for three weeks without water in winter desert conditions, tolerating huge swings in body temperature (and thus needing less water to cool itself than other mammals), and drinking 33 gallons at a time—was introduced by the Romans that humans could travel deep into the Sahara.

The experiments of dipsologists such as Adolph, thorough

though they may have been, did not measure the effects of water loss on humans beyond about 10 percent body weight, where serious physiological complications begin. Instead, for descriptions of what happens beyond that point, they had to turn to the observations of desert travelers such as McGee and people such as Pablo. In McGee's day, the old desert rats of the Southwest and Mexico had defined, in their own laymen's terms, precise stages of "desert thirst," each of which roughly corresponds, according to more modern researchers, to a 5 percent water loss by body weight. The names themselves provide a graphic progression of the stages of thirst. They run in order like this:

> Clamorous
> Cotton-mouth
> Swollen-tongue
> Shriveled-tongue
> Blood sweat
> Living death

And yet, risking all this, still wanderers and travelers head out into the deserts of Africa, Asia, the Americas, Australia, and still they die of thirst, or a dozen other ways. Each wayfarer has his or her own reason for heading out into the vast emptiness, whether it's gold, the renown of exploration, or a spiritual quest. But, at least at some level, all those reasons finally intersect the one given by Paul Bowles when he rhetorically asked himself, despite the ever-rising level of discomfort one faces in the Sahara, "Why go?":

"The answer is that when a man has been there and undergone the baptism of solitude he can't help himself. Once he has been under the spell of the vast, luminous, silent country, no other place is quite strong enough for him, no other surroundings can provide the supremely satisfying sensation of existing in the midst of something that is absolute. He will go back, whatever the cost in comfort and money, for the absolute has no price."

No *gerba* lies on the ground nearby. There is only the silence, and the camel tracks dispersing in seven or eight different directions. The first brush of morning sun skips across the tops of the orange dunes. Already you feel the air heating, although the sand on which you sit remains chilled from the night air. You dig your fingers in as if to grasp its coolness. Beside your right hand you notice a dark splotch. You look up. A *gerba* hangs from a low branch of the acacia tree. It drips one slow drop after another, and those drops are drunk, as the nomads say, by the desert.

You jump up and seize the *gerba* in your arms; then, like lifting an infant from a basket, you carefully hoist it and its leather strap from the branch. The tiny pores in the goatskin are leaking, like all *gerbas* do at times. You've watched the nomads plug the leaks with tufts of camel hair. Ignoring the pain in order to save the precious water, you yank out a few strands of your own hair, twist them, and wedge them into the hole.

The leak stops. Cradling the *gerba*, you look around again. Where have they gone? Merely off on some brief errand or another? Or do they plan to stay away a long time? You haven't a clue. What are you supposed to do? Wait here under the weak shade of the acacia tree hoping they will return and take you to safety? Or are you supposed to start walking? No matter what else happens, if you don't find them or they don't find you, your water is going to run out—sooner rather than later. Maybe they really do want you to die. Why else would they tie your hands and blindfold you? Maybe it's as simple as that. Maybe you knew too much about their guerrilla activities. Or maybe you didn't know enough. But if they really wanted to kill you, why didn't they just do it when they had the chance?

The silence gives no answers. You sink back to the sand,

cradling the precious goatskin bag on your lap. The ball of the sun has lifted a notch higher, intensifying from orange toward yellow. You shake the *gerba*, listening to its heavy contents sloshing back and forth. You guess it contains about 10 liters. As you sit there pondering the contents of the goatskin and your course of action, complex changes are already occurring in your physiology. During the night and early-morning hours since you last had a drink, your body lost about half a liter of water through its usual routes—respiration, sweat, urine. As you slowly become dehydrated, the concentration of salts and other substances in your blood plasma rises, increasing what is known as its osmolality—in essence, the amount of other substances that are dissolved in a fluid. The normal osmolality of your blood plasma ranges between 275 and 295 milliosmoles per kilogram of water. It's this varying level of osmolality that determines whether your body should open the floodgates and spill out water through your kidneys or close them down.

As you slowly lost water through the night and into the morning, your osmolality climbed above 280 and toward 290 mOsm/kg, triggering the osmoreceptors in the hypothalamus of your brain to tell your pituitary gland to release the substance that acts as the reservoir's floodgate closer—arginine vasopressin, or AVP. This signals your kidneys, which constantly filter the water content of your blood and extract urine from it, to reabsorb more of the water that they might otherwise excrete with the urine. The volume of urine your kidneys constantly produces drops as your blood plasma osmolality climbs and the pituitary releases AVP. By the time you first become aware of your thirst—which varies from person to person but occurs somewhere around an osmolality of 295 mOsm/kg—your urine flow has already fallen to 10 percent or less of its normal amount. In other words, your brain is telling your kidneys to hold onto your body's water long before you consciously sense thirst.

Sitting with the *gerba* on your lap, you've already crossed the 295 mOsm/kg threshold. You feel thirsty, but for the moment you hold off from drinking, wanting to save your water. Instead, you get up to pee. Your urine—concentrated because your kidneys are holding back water—already looks a darker yellow.

Beyond the scraggly acacia tree, the sunlight has shifted higher up the spectrum from yellow toward hot white, shifting the sand's hue along with it, and the sky has transformed from pinkish dawn to the luminous blue of full day. You search for clues the nomads may have left. You pick up the blue robe that lies on the sand. You shake it out. Nothing. You pluck up the end of the 12-foot-long blue *tagil-must*. Your spiral reporter's notebook comes tumbling out and, neatly clipped to its cardboard cover, your pen. You quickly leaf through the pages looking for a message they may have left you. The scrawled pages of handwriting it contains are all your own—interviews, landscape descriptions, odd facts. No clue from them. You slump again under the tree. Then you realize—*you have your notebook!* The notebook contains a story that will be like a sheaf of thousand-dollar bills when you finally get home! It will make your career and make you a lot of money besides. The important thing now is to keep a careful ongoing journal in it so that you can write a vivid, brilliant account of your experience on your return. The notebook, you realize, is just as important as the *gerba* full of water.

The sun has climbed another notch. You squirm closer to the acacia's trunk, where shade lingers. The sun forces you to think—hard. So what are your options? Stay here under the shade of the acacia tree and hope. Follow the camel tracks back to the last oasis and its mud village, date palms, cool ponds, and irrigation channels. Or follow one of the seven or eight tracks that lead away from the acacia tree with the hope that it will soon bring you to water, or at least to help.

Three options—that's all. Stay here . . . or go back . . . or go forward. You shake the goatskin again. Ten liters. That's worth roughly two nights of walking, or one day. The last oasis is five days' walk back. No way can you make it that far on the contents of the *gerba* alone. That leaves two options: stay put or go forward. What should you do?

Whatever you decide, you're not going anywhere until the heat of the day has passed.

Curled up against the trunk of the acacia with the *tagilmust* under your head and the *gerba* and notebook beside you, you doze on and off through the morning. The caravan, when traveling, usually didn't stop until late in the nights, and you, like the nomads, managed only a few hours of sleep before breaking camp in the mornings. You need to sleep. Each time you rouse from your slumber the air feels hotter. You must squiggle your body each time a few inches through the sand to keep in the depth of the single patch of weak shade. You know that you're saving water, in effect, simply by staying quietly in the shade. Adolph's experiments with soldiers during World War II showed that a man walking at 3.5 miles per hour in the desert sun at 100 degrees lost each hour through sweating about 1 quart of water—roughly 1 liter. Sitting quietly in the shade at 100 degrees, he lost only a quarter of that—about a cup each hour. This is one of the fundamental facts of the desert.

It is now May, and the summer daytime temperatures hit 100 degrees. You remember how at the last oasis no one moved around outside during the heat of the day. Instead the inhabitants retreated into the cool labyrinth of the *casbah*—a sort of thirteenth-century fortresslike apartment complex constructed of thick mud walls and covered passages that blocked out all sun but for small patches that entered through open skylights. Dozens of related families lived in the *casbah*, two or three hundred people, their thick wooden doors opening off the low, cavelike passages into cool, windowless rooms. It was there, while sharing heavily sugared green tea with one of the endlessly hospitable families, that you met a bright boy who spoke decent English. In late afternoon, as the sun lowered and the heat weakened, he guided you around the oasis, along the trickling and gurgling irrigation ditches, the carefully kept patches of grain, the groves of palms with their thick clusters of dates.

Under the palms, you came to a man sitting cross-legged in front of a simple mud hut who appeared to be chanting in poetic rhythms. Instead of the blue of the nomads, he wore a white turban and a patched, multicolored robe. He was thin and handsome in face and in limb, neither young nor old, neither quite dark-skinned nor quite light-skinned, and he stopped chanting when he saw you—a Westerner—walking on the path through the grove.

"Who is that man?" you asked the boy.

"He is a marabout, a holy man," the boy replied. "Sometimes he comes from far away and stays here for a time."

"May we talk with him?" you asked the boy.

"Of course," he replied. "That is why he comes here. So people may talk to him."

The marabout welcomed you with a gentle smile and gestured to a place facing him on the ground. You sat down cross-legged. With the boy translating, you explained to the marabout that you have come to the Sahara to write an article for an important magazine in America about the nomads and their struggle. He nodded and again smiled gently and noncommittally, as if the concepts of "magazine" and "America" are about as near to this particular spot as the planet Jupiter. You asked the boy to ask him if he minded telling you something about what he does.

"What would you like to know?" he asked.

You quickly tried to think up questions that would elicit information from him as well as a colorful response, snippets of quotes that you could use to give authenticity and texture to your magazine piece. Who knew? Maybe he'd even pull out a hookah and smoke a bowl of hashish with you—if these marabouts did that sort of thing. It would make a great anecdote for the article. Your editor would

love it—exotic drug use in foreign cultures was almost as good for a
travel piece as a few topless celebrities around the pool. Wherever
the interview might lead you, you decided to approach the marabout
slowly, by casting your net wide.

"What do you teach the people?" you asked.

"I am a Sufi," he replied with his quiet smile. "I teach the
traveler—the *salik*—to follow the *tariqat*, the path. I teach he who
would be a traveler on that path the seven stages of the Sufi way."

"Can you teach me about the Sufi way and its seven stages?"
you asked, now pulling your reporter's notebook from the back
pocket of your khakis. You hadn't wanted to bring it out earlier for
fear of frightening him into silence.

He didn't seem at all put off by the notebook and pen as you
quickly and as surreptitiously as possible scribbled down the Sufi
words for "traveler" and "path." You knew you could weave them
somewhere into your piece.

"You would like to learn the Sufi path?" he asked, again with his
quiet smile.

"I am very eager to learn," you said.

"You would like to learn to be a traveler?"

"Yes, I would," you said as sincerely as you could.

It was partly true, anyway. You'd long been attracted to Eastern
mysticism, and you'd always thought that someday you'd follow that
path. Not that you knew that Sufism was at all the same path. Nor
was this the time and place to follow it. What you were really here
for was to learn all you could from the Sufi that would make good
material for your article.

It was at this point that the Sufi asked if you knew of the great
Sufi poet Hafiz, who lived in Persia in the fourteenth-century.

"I don't," you said, your blue felt-tipped pen at the ready over
the open page of your notebook. "But could you tell me how to spell
his name?"

It was then that the Sufi reached over gently and took the notebook from your lap. Laughing, he looked up at the palm-fringed

desert sky, whirling the notebook in one hand over his head, and in a loud, rich, melodious voice chanted out a verse of Hafiz:

> Oh thou who are trying to learn the marvel of love
> from the copybook of reason,
> I am very much afraid that you will never really see
> the point.[2]

Under the reddish sky of sunset you pour another long drink from the *gerba* down your dry throat. Through the hot afternoon, dozing under the acacia tree, you managed to hold yourself to one long drink. Adolph and fellow researchers determined that rationing one's water does not help in the desert—that whether you drink quickly or drink slowly, your body finally needs the same quantity of water. In some ways it's better to drink sooner rather than later, they found, because without drinking—no matter how hardened you think you are to deprivation—you cannot maintain the strength to march. Still, as you sit under the acacia, you feel better practicing some restraint.

Each long drink amounted to about 0.5 liter, now leaving you 9 liters in the *gerba*. But during the heat of the day you lost one cup—about 0.25 liter—each hour in sweat to cool your body in the 100-degree heat even in the shade of the acacia. So in the course of the day you've drunk 1 liter but sweated out 2, on top of the 0.5 liter you already were down on waking at dawn. That leaves you now down a total of 1.5 liters, or just over 2 percent of your body weight— still not quite enough to impair your efficiency, but certainly enough to make your body want water.

Paradoxically, if you had all the water in the world to drink, you

---

2. Quotations and translation from *The Persian Sufis* by Cyprian Rice (London: George Allen and Unwin, 1964).

could kill yourself with that, too. Over the years, some fifteen marathon runners have died by drinking too much water during races—a condition known as water intoxication. The gut, with its blood diverted to the muscles, cannot absorb the excess water until the race's end, when it floods the bloodstream, upsetting the body's sodium balance. The same upset of the sodium balance can occur to workers in the desert heat, even if they drink water copiously and regularly. The body can sweat out its sodium—known as an electrolyte, due to its role in transmitting electrical impulses in the body— that normally is contained in the blood plasma and other fluids. Although the water lost from sweating is replaced by drinking, the sodium, normally ingested by humans in the form of salt in their food, is not. This loss causes the osmolality—the amount of a substance dissolved in a fluid—of the blood plasma to drop in relation to the osmolality of the water inside the body's cells, and the body tries to equalize this difference in osmotic pressure by sending water from the blood plasma through the cell membranes into the cells. This can cause the brain cells to swell, resulting in seizures and, if untreated, death. For these reasons, "sports drinks" advertise that they contain electrolytes to replace those the athlete loses in sweating.

But the water in your *gerba* contains small amounts of salt from the wells of the oasis and from the goatskin in which it rides. Sodium depletion is not a problem. Furthermore, in the cool of the evening, desert wanderers lacking water often find that their sense of thirst diminishes. You now feel more clear-headed than you did during the heat of the day. All afternoon you've been reviewing the possibilities. It's now become obvious that the nomads didn't go off on some brief errand with the intent to return. You begin to think maybe you should move on. You *know* you're supposed to stay put. That's what the wilderness manuals and camping handbooks say. The first rule about being lost is, don't get any more lost. Stay put and wait for your rescuers. But who's going to rescue you? The only people aware that you're sitting under this lone acacia tree in the middle of the Sahara Desert are the nomad guerrillas, and their entire intention might be to make sure you *stay* lost. Of course, there's your editor. You dropped him a postcard two weeks ago telling him about your plan to head out with the caravan, but it will be weeks more before he receives it and still weeks more before he misses you. Besides, what's he going to do then? Cancel a month of luncheon dates, buy a designer safari suit, and come to the Sahara to search for you? Not likely. Or maybe he'd consider it if he thought there might be some publicity in it for the magazine and his own editorial savoir-faire. But by then the supposed object of his dashing mission—you—would be long dead.

Yes, maybe you should move on. Choose one set of camel tracks, follow them, and hope for the best. Maybe wait another hour or two to see if they return this evening, and then move on. It's a big, scary thought—especially the thought of the desert by day, of leaving the safety of the acacia's tiny patch of shade and walking alone into that huge burning emptiness with zero idea of where you are heading. But if you don't walk into it—if you don't face it—that huge burning emptiness will just as surely kill you where you sit.

There is a brilliant half-moon high in the black desert sky as you prepare to set out. The air temperature has dropped to a soothing, almost chilly 70 degrees. On the dry skin of your face, however, you can still feel the invisible infrared rays of the day's heat radiating from the sand beyond the acacia tree like the sudden heat from the orange coils of a broiler oven that's been opened on the far side of a kitchen.

You gather up the robe and slip it over your head, pushing your arms into its flowing sleeves. You attempt to wrap the *tagilmust* around your head in the nomad fashion, but it unravels each time. Finally you simply give it several turns around your waist and knot it. You slip your reporter's notebook into the back pocket of your khakis. At times during the day you made notes about your situation and your thoughts. It is all more material for your article.

You sling the *gerba* over your shoulder by its leather strap; it and the notebook are the twin safeholds of your future. The high half-moon illuminates the sands around you like wide, silvery swells. In the moonlight, the camel tracks radiate outward from your lone aca-cia tree in dark lines like the threads of a spider's web. Each set of tracks seems to mark where one of the eight nomads and the two or three camels he lead headed in a different direction. You've been pondering this problem since sunset and finally decided to follow the set of two tracks intertwined together that are heading east. That's deeper into the heart of the desert, but it's also the direction in which the caravan had been bearing before they abandoned you. Since the last oasis where you'd met the marabout they've traveled five days without water. In the next two or three days, the camels will need to drink and the nomads will need to refill their empty *ger-bas*. They'll have to find a water source soon. So runs your logic.

You step out from the acacia's small patch of moon shadow and set one sandaled foot and then the other on the open desert, like stepping gingerly out of the controlled comfort of a space capsule onto the waterless, airless surface of the moon. You are leaving be-hind your only tie to the fertile, green world into which you were born. A sudden tremor of anxiety jumps up to your chest. The no-mads might be returning any hour. You will be gone. You will have missed them. But then you realize that if they wish to find you, they'll track you easily. They are born trackers. They can squat down beside a set of camel tracks and, like a fingerprint expert, identify by the particular marks it leaves in the sand an individual animal they haven't seen for years. They can read by a person's footprints how fast he is traveling, what condition he is in, whether he is in a hurry due to some suspicious activity. "A man's conscience," the nomads say, "is reflected in his tracks."

"The desert and the track cannot lie," they say.

And they have another saying—one on which you're about to stake your life: "The track brings the man."

You don't care if their track brings you to them or to someone else, or your track brings them or someone else to you, as long as the meeting results in copious draughts of water to pour down your throat.

At first the walking is easy, with the broad, thin desert sandals you purchased back in the last oasis staying atop the soft surface. The dunes roll gently up and down, like great ripples in the shallows of a beach. These great seas of sand cover only a portion of the Sahara and are known in Arabic as *erg* or *areg* ("sand deserts") or in Tamachek as *edeyen* where the dunes are flat and *iguidi* where they are steep. Much of the rest of the Sahara is covered by gravelly or stony plains known as *reg* or *serir* and by very rugged rocky ones called *hammada*—which means, with very good reason, "to die off."

Fortunately for you, in the dunes of the *erg,* it is easy to follow tracks. You trace the dark intertwined thread of the camel tracks over their silvery swells. Your thirst seems to have diminished. You feel strong. You like the cool watery heft of the fat *gerba* under your arm and the thick notebook full of possibility stuffed in your back pocket.

After waving it over his head and quoting the Persian Sufi poet Hafiz on the futility of learning love from books, the Sufi handed your notebook back to you.

"You may keep writing in your small book if you like," he said.

As the sun moved down through the desert sky above the shading palms, you slowly learned the precepts of the Sufi way by asking him your careful questions, through lines of poetry the Sufi quoted with his chanting, melodious voice, through his pithy bits of explanation. You understood how, founded in Persia in the ninth century A.D., Sufism is part of Islam but also a reaction to Islam's austerity and how, through poetry and music, dancing and chanting, it aimed at abandoning the self for a mystical, ecstatic union with God.

"To be a Sufi," he said, quoting Abu Said of Mihneh, who lived in the eleventh century, "is to give up all worries, and there is no worse worry than your you-ness. When occupied with self, you are separated from God. The way to God is but one step: the step out of yourself."

You scribbled as quickly as you could in your notebook to take down all the quote. Yes, you thought, this brief encounter with the Sufi would make rich color for your magazine article. Your editor had never minded a little Eastern mysticism thrown into a travel piece, especially if it contained the suggestion of ritualized sex, as long as he didn't have to think too hard about the concepts. While it didn't appear that ritualized sex was up the Sufi line, trials and trances certainly were, and those weren't bad for color, either.

"The seven stages," the Sufi went on in his own words, "come not through *mushahada*, contemplation, but from *mujahada*, striving. The traveler must transcend the self by constantly training the purity of his intentions and eliminating all *gharad*, ulterior motives."

The Sufi paused. You kept writing.

"Do you understand?" he asked.

"I do," you replied, looking up quickly from your notes.

"And you are still wishing to know the Sufi way?" he asked.

"I am," you said. You still needed more material for the piece. Maybe he'd eventually bring out the hashish or go into a trance. "Let's continue."

"The traveler on the Sufi path must know that God Himself will help him reach his goal of abandoning the self by sending the traveler mysterious afflictions," the Sufi said. "This will help the traveler attain *maut i ikhtiari*—voluntary death."

You are awakened by heat. Burning desert heat. On your head, your arms, your legs, inside your mouth and throat, even in your lungs. You struggle to sit up. You're at the apex of a small dune. This is the spot where you laid down in the predawn darkness to rest. You'd followed the intertwined camel tracks for seven hours or so, covering—you guessed—between 15 and 20 miles across the *erg*. As Adolph's calculations predicted, the effort to cover that distance— 20 miles per gallon in the cool of a night march—has sucked 3 liters of water from your body. Wisely, you stopped to drink four times, about 0.5 liter each, even though you didn't feel you were that thirsty. Humans working in the heat, physiologists have found, never have the urge to drink as much fluid as they've lost until hours later. Your periodic water breaks replaced 2 of those 3 sweated-out liters. But overall you lost 1 liter during your night's trek on top of the 1.5-liter deficit you had before you even started walking. So, as you laid down in the predawn darkness to rest atop the sand dune, you were experiencing a 2.5-liter deficit, and lost another 0.5 liter in the four hours you've been asleep in the early sun. As you awaken in the burning heat, you're down 3 liters, or (figuring the weight of 1 liter of water at 1 kilogram and your normal body weight at 70 kilograms or 155 pounds) a total water deficit of 4.3 percent of your body weight. This is still close enough to that 3 to 4 percent range where your body experiences only moderate impairment of efficiency, but you're well into that first phase of desert thirst that the old desert rats and prospectors called "clamorous."

You're extremely thirsty, especially in the late-morning sun. Your mouth feels dry. Each breath seems like a hot wind has entered your lungs, leaching the moisture from them. The sunlight strikes your naked forearms, protruding from the sleeves of your robe, with an almost visceral sense of pressure, as if brushed by the breath of a steaming teapot. In the cool of the night you have forgotten the sun's absolute power during the day. It now charges the air and sand around you with its force, as if the molecules themselves begin to vibrate.

You grab the *gerba*. Before you can stop yourself, you've drained an entire gurgling liter of its contents down your burning throat. The cool, dung-flecked water instantaneously gives you some relief. You carefully retie the thong around the goatskin's neck. It now feels noticeably smaller and lighter than when you began—6 liters sloshing around instead of 10. You cradle the goatskin in your lap. You want to drink more, but you don't dare. You have no idea how far you must walk to find a well, or an oasis, or even another person, and the contents of the goatskin is the only thing that will get you there.

Looking out from the top of the dune, you realize that every square inch of barren sand as far as you can see is seared by the same intense focus of the sun's rays. It's like you've opened the door and stepped into a sauna as large as a continent. How far must you walk even to find a simple spot of shade? Pulling yourself to your feet, you stagger slightly—already the lack of water has affected your balance—and try to wrap the *tagilmust* around your head for protection from the sun. You manage to tie a kind of topknot and wrap a dangling end around your nose and mouth with only your eyes showing. The arrangement seems to insulate your head from the sun and cool the air slightly as you breathe it through the veil. Now you understand why the nomads adopted this dress to cover as much surface area of their body as possible from the desert sun. Adolph's studies found repeatedly and surprisingly that a man wearing a uniform and sitting in the desert sun saved over 0.25 liter of water *per hour* in sweat compared to a man sitting in the sun nearly nude even though he might feel more comfortable without the clothing.

You set out again, slowly, wrapped but for your eyes, along the camel tracks. Summer is coming to the Sahara; this, by far, is the hottest day you've experienced since the caravan entered the desert two weeks before. You have read that on a summer day the Sahara's air temperature can reach 120 degrees in places, and the sands heat up much hotter than that; the air has a bone-dry humidity of 5 percent. As you walk, the sun's rays drum down like a pounding rain. The air shimmers and snaps with them. The heat is a viscous substance you must push your body through. The blue robe helps shield you, but you feel the heat rising through the thin leather soles of your sandals. The surface of your corneas feels dry and hot, as if you stand before a fire. The water has helped, but not for long.

On foot, in the hot desert sun, the human body sweats out water to cool itself at an incredible rate—a liter or more per hour. After an hour's struggle through the dunes, you've lost another 1.5 liters, bringing your total water deficit to 3.5 liters, or a deficit of 5 percent of your body weight. You're down only about a gallon out of the 10 or so gallons your body normally holds, but already you're entering the old desert rats' "cotton-mouth" stage of thirst. You become increasingly weary and start to stumble. Your pulse rate climbs from 70 beats per minute to 100; as you dehydrate your blood volume shrinks, and your heart must beat faster to keep enough blood flowing to body and brain. Your skin feels flushed, and you sense the beginning of a tingling sensation beneath your robe, the first sign of the numbness that afflicts victims of thirst. Dehydration can cause a sense of restlessness that may be related to low blood pressure. A victim of dehydration often feels less discomfort by keeping moving. You push on.

You begin to think more and more about water. You fantasize about water—water sloshing around in the *gerba*, water trickling in cool rivulets under palm trees, water in blue swimming pools, water arcing up from chilled drinking fountains, the glasses of water you've spilled down a bathroom sink after brushing your teeth and rinsing your mouth. If you could only take them all back now! By force of will you don't drink from the *gerba*. You don't know how much longer you can hold back. You keep pushing yourself forward into the hot desert sun. The saliva in your mouth feels sticky and your tongue sometimes sticks to the roof of your mouth, taking a conscious muscular effort to pry it loose. Your breath comes in shorter gasps, and the air you suck in feels hot. The lining of your drying throat can no longer provide the moisture to humidify and cool each inhalation. A lump rises in your throat that you cannot swallow away; the sides of your trachea feel as if they stick together. Your face feels full as your facial tissues shrink and stretch across your cheeks and jawbone. Periodically you hear a drumming and snapping sound—the drying of tissues in your ears.

Your pace slows. Over the second hour you lose another liter of water, for a total deficit of 4.5 liters, or 6.4 percent of your body weight. The dunes look white-hot in the midday sun. Each footstep demands greater effort than the last, each a conscious act of will— *lift . . . place . . . lift . . . place.* If you were traveling with a partner, this is about the point where you would start to argue, become suspicious of each other, even separate as you trudge over the desert sands, then rejoin simply for the sheer satisfaction of arguing again in your cracked, husky voices. But you have no partner. Instead your mind targets your editor. He's ensconced in a leather booth in some Manhattan bistro with the waiter pouring out a bubbling, clinking stream of ice water into his glass. He glances at the menu just long enough to order a poached filet of wild Copper River salmon bathed in a pool of dill-cream sauce and a chilled bottle of Vouvray, then goes back to lamenting across the table to his lunch partner about the sorry state of the magazine industry. When he commissioned the piece, did he have any idea what you would face? Did he care? Did he actually want the story, or only want to have a look at what you came back with? And if it didn't appear all that compelling a story, would he toss you the standard 25 percent kill fee and be done with it? And if you didn't come back *at all*, you now think as you slog through the burning sands, feverishly cogitating on your editor's demonic logic, *he'd even save himself the 25 percent!*

In a hollow between dunes you see a heap of dung. The two camels stopped here. Their tracks move about in ambling circles. You see footprints here, too. You squat down and study the state of the dung's dryness, crumbling it between your fingers. A nomad could tell precisely when the camel had been here—the dung is totally dry after two days. The best you can do, however, is guess that it's less than two days but more than a few hours old.

You stand up, staggering dizzily. You try to make out the intertwined camel tracks again. But the two camel tracks split. One veers south across the next dune. The other keeps to the east. There must have been a rider on each of these camels. Which should you follow? The wall of an insurmountable problem now looms before you. You decide to sit on the hot sand to think it over. Your legs suddenly crumple underneath you, and you realize right then that you're not going to walk any farther this day.

You set out again in the evening. You've survived the blazing heat of the afternoon by scraping away the hot surface of sand to the cooler layer underneath and sitting on it, using your robe as a kind of tent around your body. Adolph's studies revealed—as nomads and old desert rats surely knew intuitively—that the best means of saving your body water in desert heat of 100 degrees was to sit quietly in one's clothes in good shade or under a night sky, both acts using 0.3 liter of water per hour in sweat. Sitting clothed in the sun jumped the sweat loss to 0.5 liter per hour, and lying clothed in the sun, which exposes more of your body's surface to heat gain from the sun's rays, pushed it to around 0.7 liter per hour sweat loss. Walking nude in the sun was the worst thing the desert wanderer could do— a whopping 1.2 liters loss per hour.

Even in your tent of shade your body had to cool itself, squeezing another liter of water from its sweat glands that instantly evaporated in the 5 percent humidity. You had to drink—a lot. Two liters. This doesn't nearly make up for your deficit. You're still down a total of 3.5 liters—5 percent of your body weight—but you're feeling much stronger than you did in the daytime. The goatskin hangs flaccidly in your arms when you pick it up. You shake it. Only 4 of its original 10 liters still slosh about. You have no idea how far those 4 liters will have to take you. You realize how deeply you depleted your water reserves—both in your *gerba* and in your body—by trying to walk in the heat of the day. When you pee, the small stream of urine looks a dark orange. Your kidneys hold back all the water they can.

The half-moon, tonight closer to a three-quarters moon, is high overhead and very bright in the clear desert air. You follow the camel track that bears east instead of the one heading south. That's the way the caravan was headed to start, and you stick to your conviction that that is where the next water must lie. Or are they testing you somehow? Is the split in the path a trick? But you know they'll need water in the next day or so. If you can do another 15 or so miles tonight, and another 10 or 15 tomorrow or tomorrow night, you're bound to reach the watering spot. But can you walk that far on 4 liters?

The dunes roll on unchanging in the sand sea of the *erg*. You carry on, *gerba* securely tucked under arm and notebook stuffed in back pocket. During the heat of the afternoon you even managed to make some notes on the Sufi belief that they are travelers on a quest. In the bright moonlight you trace with your eye the dark thread of camel tracks rolling over the dune tops and down into their shadowy hollows, on and on, until they disappear where the black sky rests on the desert horizon. To your mind, it's no accident that Sufism was founded in and around deserts. In the barrenmost heart of the desert, the central human metaphor is the journey—the journey to come out on the other side, alive, but somehow transformed.

Sitting with the Sufi back at the oasis, the blue of the late afternoon sky turned to the pinkish orange of evening. You pulled a few faded, threadbare bills of the local currency from your pocket, and asked the boy who served as your translator to bring some green tea and palm sugar for the Sufi as a gift. He returned a few minutes later with two bundles wrapped in dried straw and a fat melon. When you offered them to the Sufi, however, he declined them.

"No, they are for you," the Sufi said. "You eat the melon and drink the tea. The Sufi traveling along the path to love must pass through the stage of poverty. When the self that once desired riches no longer exists, the stage of poverty is transcended, and the beloved is all that matters. That is why we say *idha tamima'l faqr, fahwallah* 'where poverty is complete, there is God.' "

Now you understood the Sufi's patched robe and simple hut. Their original outfit centuries ago had been a coarse robe of undyed wool called a *suf*—hence the name Sufi.

The boy ran with the gifts to a mud hut nearby, and a woman there prepared the tea in a pot on a small brazier full of glowing coals and sliced the melon. When the glasses of tea and slices of melon were brought to you on a small wooden tray, the Sufi, with his quiet smile, gestured for you to eat and drink.

You did, gratefully. It wasn't as if you'd taken the vow of poverty yourself. The melon was sweet and juicy and the tea sweet and hot, and though it was thick with tea leaves and bits of fiber from the brown palm sugar, it was curiously refreshing, as ever, in the desert air.

"What are the other six stages?" you asked the Sufi as you sipped tea and ate melon, trying to keep your notebook and pen at the ready. This presented some logistical problems, because by tradition in this part of the world you could use only your right hand to eat, never your left, and your right was also your note-taking hand.

"We Sufis have no single doctrine," said the Sufi. "Each *murshid*—guide—might have his own set of stages, but many of the stages are the same."

He listed the most common ones:

      1. *Tawba*—repentance or conversion
      2. *Wara*—fear of the Lord

3. *Zuhd*—detachment
4. *Faqr*—poverty
5. *Sabr*—patience or endurance
6. *Tawakkul*—trust or self-surrender to God
7. *Rida*—contentment

Then, in his melodious, chanting voice, the Sufi looked toward the sky and quoted the Sheikh Attar ("the Druggist," for that had formerly been his profession in thirteenth-century Persia), one of the greatest writers and poets of the Sufi mystics, who had expressed his version of the stages this way:

There are seven valleys on the way. . . . The first valley is questing and seeking. Next to it is the vale of Love; then the vale of Knowledge. The fourth is detachment and liberty of heart. The fifth pure unification. The sixth grievous bewilderment. The seventh, Poverty and utter loss of self. After that valley there is no more deliberate advance: going is forgone, henceforth one is drawn.

You sit and rest for a few hours before dawn, dozing fitfully with head on knees. As the eastern sky brightens, you stand up stiffly and move on. Your plan is to travel as far as possible today before the sun heats the desert sand. You figured during the night's march you logged nearly 15 miles again. But the *gerba* now feels like an empty sack. You shake it. You're down to a mere 2 liters. You walked slowly, breathing easily, to avoid sweating or losing extra moisture through too-fast respiration, but even in the cool night air, you had to drink 2 liters to replace what you were losing and simply to keep up your strength to walk. You've maintained your level of dehydration where it was when you started the night's walk—a 5 percent water deficit by body weight. But now you have the heat of a desert day ahead of you, a mere 2 liters left in the *gerba*, and no idea how far you must walk.

After the first hour, the terrain begins to change. In the hollows between the rolling soft dunes you walk over a gravelly, claylike hardpan. Then the dunes suddenly end—you must have left the *erg*, at least for a while. You take this to be a good sign. Instead you're on a rolling, gravelly plain—the *reg*—although it's still soft enough to show the camel tracks. Half an hour later you reach what appears to be an old, dried-up streambed. A few tufts of vegetation cling to its gravelly crease. The tracks keep going. Another half hour across the plain you reach another shallow, dried-up channel. Here grow two stunted acacia trees. There must be water deep underneath the surface. The sun has now climbed high in the morning sky, and the desert floor has begun to bake. You know you should stay under the shade of the acacia until the heat of the day has passed. But you have only 2 liters of water left—the thinnest of cushions against the intensity of the desert heat. You don't know if you can last another whole day waiting here plus another night's march on the sloshing puddle of water that remains in the *gerba*. You sense that water lies close ahead. You know this will be a fateful decision.

You decide to keep walking.

By noon you know you are in trouble. It doesn't matter anymore whether the decision to keep walking was right or wrong. What matters to you now is simply to keep going. Walking through the hot morning sun, you've drained the last two liters. The *gerba* dangles empty from your shoulder. But the water you drank through the morning hasn't replaced the 1.5 liters of water you sweated out during the morning's hike. Added to your previous 3.5-liter deficit at dawn, this puts you down 5 liters, or 7.1 percent of your body weight, back into the cotton-mouth range. With every effortful step over the gentle rises of the gravelly plain of the *reg* you expect to see a caravan stopped around a well, or the deep green palm fronds of an oasis tracing along some watercourse. You see neither of these things. What you do see is a bluish haze in the distance that looks like mountains.

You can only think about following the camel tracks—their slow, purposeful plod—and about the visions of water you'll find at the end: the oasis, the swimming pool, the drinking fountain, the glass of drinking water at your editor's lunch table. You force yourself to keep walking. You are now in a sprinting race—a hobbled, stumbling sprint—between the rapidly dwindling moisture in your body and the distance that lies to relief. In an hour of plodding in the desert sun you've lost another liter. Your body starts moving more quickly through the phases of dehydration. You're now down 6 liters—8.6 percent of your body weight. At an 8 percent deficit, as Adolph's experiments found with soldiers, your heart is pounding 40 beats per minute faster than normal—or 110 beats instead of 70— but it's putting out considerably less blood volume with each beat than normal. As the water in your blood plasma leaches away to cool your body in the form of sweat, your blood is thickening like oil. Your pulse rises and stroke volume drops, apparently because it takes your heart more and more energy to pump that ever-thickening blood.

If you looked back at your own tracks, you'd see them weaving and stumbling across the *reg*. Still you keep following the camel tracks toward the bluish mountains, past 10 percent deficit, 7 liters down, and you stumble onward into the "shriveled-tongue" phase. As your blood thickens, the process of osmosis causes water from inside your body's cells to pass through the cell walls and help thin your blood plasma. In effect, your body sucks the moisture out of its living cells to keep its blood liquid enough to flow.

With each beat you hear a crackling in your desiccated ears and see stars jump before your eyes. At times the noise sounds like music coming from just over the next rise. You try to walk faster to reach it, but your feet drag as if the dry white stones themselves were tied to your sandals, and then you realize once again it's only noise in your ears. Your skin turns numb and tingly. The khaki pants you wear beneath the robe grate annoyingly on your skin. You have the robe; you don't need the pants. You lift your robe, unfasten your belt, drop your khakis down your legs, followed by your boxer shorts, and kick them off on the gravelly plain. The freeness around your legs gives some relief from the grating of your leathery skin. But you see you've left your notebook in your pants pocket. You pick it up. You have no pocket for it in your T-shirt beneath the robe. It's too much to carry both the *gerba* and the notebook. You don't need the empty *gerba*. You slip it from your shoulder, drop it to the sand, and stumble onward.

Your head now throbs as if an iron band tightened across your skull. You scratch and tug beneath the turban at your pounding head. You want to unwind it, throw it away. But some deep instinct tells you to keep it on. You're now breathing heavily through desiccated airways. Your eyelids have stiffened from the dryness and don't close properly over your eyeballs. You're developing the desert-thirst victim's characteristic "winkless stare." Your numbed limbs no longer seem attached to your own body; the hand holding the notebook drifts oddly out in front of you, as if carried by someone else. Late in the "shriveled-tongue" stage, reported McGee, one desert prospector, "seeing a luscious-looking arm near by, . . . seized it and mumbled it with his mouth, and greedily sought to suck the blood; he had a vague sense of protest by the owner of the arm, who seemed a long way off, and was astounded two days later to find that the wounds were inflicted on himself."

Your moist tissues exposed to the air dessicate first—mouth, nose, lips. Your body has literally begun to mummify from the outside in while you are still alive. The saliva now has entirely ceased to flow in your mouth, and the mucus has dried into a thick film over your tongue, deadening it. The gums of your teeth and the linings of your nostrils, ordinarily containing much moisture, shrink as the water leaves them. Your nose is literally growing smaller. The lump in your dried-out throat feels like a solid, immovable rock, despite your constant and involuntary reflex to swallow. Finally your tongue hardens into a dead weight. You notice an odd clunking sound in your mouth. Then you realize that with each of your slogging footsteps across the baking *reg*, your hardened tongue swings like a pendulum on the still-moist tissues of its root and clunks against your teeth. It is this that gives the "shriveled-tongue" phase its distinctive name.

In this phase—if not before—thirst victims often will drink almost anything to wet their mouths. *Mariposia* is the name given to the drinking of seawater, *hemoposia* to the drinking of blood, and *uriposia* to the drinking of urine—either their own or an animal's. Uriposia is done commonly by those in a bad state of dehydration and was a well-accepted practice in the Southwest and Mexican deserts during McGee's day, at least to wet one's mouth with urine. Even when dehydrated by 7 or 8 percent of body weight, the human body, which normally produces about 1.5 liters of urine per day, continues to produce urine at a reduced rate, although it is believed that the flow ceases altogether as one approaches lethal water deficits. Whether drinking urine helps is not entirely clear, and may depend on many factors; drinking seawater does not slake thirst due to its high salt content. Animal fluids apparently are a different matter. Desert nomads sometimes sacrifice their camels when in desperate circumstances and, piercing the abdomen with knife or sword, drink the bloody, yellowish green liquid—vegetal pap—that pours forth from the incision. One U.S. calvary detachment chasing a band of Indians in 1877 in Texas promptly ran out of water, and members reported that they survived only by sucking on the coagulated blood of their dying horses as the party staggered back across the plains to the lake where they'd begun their intemperate pursuit.

Your tongue clunking against your teeth, you stop. You feel a distant need to pee. If you could just moisten your tongue enough, you think, you could keep going farther across the plain, toward the bluish mountains. You lift your robe. Swaying in the desert heat, you try to pee. It takes a long, long time to will up the urine from deep inside you. Your eyes seem to dry further as you stand there, trying to concentrate. It's difficult to see the camel tracks except as a blurry line heading toward the bluish mountains of the horizon. Finally, stinging, the urine trickles out—the color of rusty water. You catch an ounce or two in your cupped hand. You bring it up to your mouth. There is no sense of disgust, only a desire for its liquid. You suck it between your thin, dried lips and hold it in your mouth to moisten your tongue—it tastes salty and wet—and swallow it down your throat, momentarily alleviating the lump. You hold out your hand for more, but nothing comes, no matter how hard you try.

The first blast of wind strikes you from behind, knocking you forward as if you'd tripped on a large stone. The second gust knocks you off your feet. You're on hands and knees on the baked sand and stones, your reporter's notebook still clutched in your right hand, as the sandstorm sweeps over you in a hot, tannish yellow cloud of dust and fine particles of sand. Crouched on the desert floor, the wind whipping your robe, the sand stinging your skin, you try to wrap your *tagilmust* tightly around your face, but the wind rips its end from your hands so that it flutters like a flag in a strong wind. Instead, you pull your head down into the robe, out of the stinging sand and hot wind, and sit there, unsteadily, feverish, only wanting to go on.

You remember the first time you saw the Sahara; it gave you the idea and the desire to come back to follow the nomads. You'd been in southern Spain on an assignment, and you'd taken a ferry to Morocco, rented a car in Tangier, and drove it south, over the mountain range known as the High Atlas, until you reached the edge of the desert. As you crested the mountain pass on the narrow paved road, you could see far out across the great expanse of the Sahara. Far down below you a sandstorm was blowing, filling the great basin of the desert with a luminous yellow haze like an ocean, the sky still a crystalline blue above. As you drove down into it toward an oasis on the desert's edge, the storm and its blowing sand literally blasted the paint off the front of the car.

You are now at the bottom of that blasting yellow ocean, huddled beneath your robe, arms wrapped around you, cradling your life.

The storm ends just as suddenly as it arose. You push your head up through the robe. It takes you three tries to get to your feet. You try to get your bearings. Something is missing: your reporter's notebook. You were holding it in your hand, but you had to set it down to create a shelter out of your robe. It is now no doubt blowing at 40 to 50 miles per hour somewhere to the southeast.

For the moment, you're too dehydrated to understand what you've lost. You stagger forward, following the dim line that remains of the camel tracks.

At a 12 percent water deficit, you've crossed the divide. You can no longer swallow on your own, and if you do find water, you will need someone's help to rehydrate. The potential death zone of dehydration begins soon—at about a 15 percent deficit. No one knows exactly where the boundaries lie that mark the last two phases of thirst described by McGee. The lips and gums and tongue and other tissues crack open with festering wounds, and the tongue swells and forces its way between the teeth and out of the mouth. Flies gather on the tongue and other tissues and drink the blood and fluids that weep from the cracks and the tears of blood that fall from around the eyes. Carrion-eating birds congregate around the victim. A friend of McGee's, an old hand in Mexico's Seri Desert, rescued a vaquero who had been thrown by his horse three days earlier, and graphically described the victim when he found him as "sweating blood and fighting buzzards." For this stage of thirst, McGee used the term "blood sweat."

The final phase—"living death," in McGee's parlance—is a continuation of the previous stage. If the trails the victims leave in the sand and dust are followed, he wrote, one can see that they rip at their scalps and tear out clumps of their hair to lessen the sense of a band tightening around their skulls. They hallucinate that rocks or shrubs or sharp-thorned cacti are pools and carafes of water in which they dig their fingers or onto which they skewer their faces in the attempt to take a drink. Even to more modern researchers, it is not clear what finally kills a victim of dehydration. In cases of desert thirst in hot temperatures, heatstroke may be what kills the victim, as a severely dehydrated victim cannot effectively dispel body heat.

But even in the latter stages, victims of thirst are able to follow a trail to water even though they can barely move or see. Something tells you to keep moving forward. Now you're stumbling badly. You constantly hear noises and music and the harsh rasping sound of your own breathing through desiccated lungs. The desert periodically blurs to a luminous hemisphere above and a darker hemisphere below bisected by the camel track. The robe grates against your parched skin, and you feel a painful tightening around your chest. You don't need the robe. It's too hot. You pull the robe over your head, strip off your T-shirt, tug off your *tagilmust* with its iron bands that have tightened around your skull. The naked legs and arms, drying and cracking in the sun, are not yours. They were left here in the desert to turn to leather by some passing animal. There's a person here, too. You watch him from a distance. He's very tired and thirsty. You leave him behind, too, along with the pile of clothing and the arms and legs.

You stumble onward, following the thin line that leads between the blurry hemispheres of light and dark.

Late that afternoon, a young nomad boy is rounding up his family's camels and goats that have been grazing on a few tufts of grass in a *wadi*—a seasonal riverbed—when he sees a strange naked form crawling along the dry stones. At first he thinks it is some unknown animal—perhaps one of the crocodiles he has heard about that live in the lakes far to the south—but as he approaches cautiously, he sees it resembles a man. Not Imazaghen but a French—the nomad name for all Europeans. His skin is a peculiar gray, his lips have shrunken and turned black, his nose looks like two round holes in his skull, and his eyes, set deep in their sockets, stare at the boy unblinking while the strange man-animal's chest heaves in and out with a bellowing sound. The boy nears the man-animal, his staff at the ready to beat him away if necessary, but the man-animal seems to recognize a living form; his forearms crumple, and he collapses face first into the sand and rocks.

The boy leaves him there. He runs in sandals and robe back to the nomad encampment in the small oasis at the foot of the blue mountains and tells his father what he has seen. Soon a group of nomad men ride camels quickly to the spot, led by the boy riding in front of his father. They know it is not a crocodile or a man-beast but a desiccated French. They pick him up—he is not very heavy, like a dried-out goatskin—prop him into a saddle, where he is held securely by the nomad sitting behind him, and ride him back to their camp.

They lay him on the sandy earth, and the nomad women—some of them bare-breasted—come from their low, goat-hair tents to look at the dried-out French as he is lifted down from the camel. His skin is shrunken around his bones, his belly a deep hollow.

"Like a baby goat that has lost its mother," one of the women says.

If this were the emergency room of a Western hospital instead of a remote Saharan oasis, medical personnel would immediately fix intravenous drips to the victim that contain mostly water, adjusting their exact content once an imbalance of sodium or other substances in his blood was determined. But the nearest medical care—a single village nurse—lies on the other side of hundreds of naked miles of *reg* and *erg*. Together, the men and women make preparations for the traditional method to revive desiccated victims of desert thirst, the method that Saharan dwellers have developed over centuries of experience, as recorded in detail in the 1930s by a French military doctor among the Goranes of the Central Sahara. Some victims can no longer swallow, and if the victim is still able to drink when he or she is found, the desire is to pour water down the throat without stopping until vomiting occurs.

The women now fetch goatskins of water and a robe. The men squat around his naked form and would have stripped away any clothing had he still worn it. The victim is semiconscious, stirring vaguely, mouth moving as if trying to talk. They very carefully pour drops of water onto certain areas of his skin—into the hollows formed atop the shoulder by the clavicle, the armpits, the bend of the elbows, the hollow of the stomach, the folds of the groin. The nomads pat the moistened areas lightly with their hands. They slowly pour out water—drop by drop—over the victim's head and face and pat them lightly, and then repeat the sprinkling and patting on the victim's legs. They take the robe, sprinkle it with water, and cover the moistened victim with it. He falls into a light sleep. A short while later, he wakes. He makes noises with his mouth. Splashing the contents from a *gerba* into a wooden bowl, the nomads give him a little water to drink. He swallows with great difficulty. They then remove the robe and again drip water on all the sensitive areas of his skin, pat it, cover him again, and give him a small drink, repeating the process over and over. Finally the victim falls into a deep sleep.

What you first hear as you come to consciousness is music—a slow, rhythmic beat of drums and what sounds like tambourines. There are voices, too—chanting voices. You open your eyes. You are lying on your back on the ground. The sky vaults overhead in the cool purples of dusk. Palm fronds fringe the purple sky, and orange firelight flickers up their trunks, illuminating their fat bunches of dates. Blue-turbaned heads pulse rhythmically over you, a circle of heads that constricts around you in a ring, rhythmically withdraws to the clanging, thumping beat, then constricts again, slowly turning around you like a wheel around its axle. The nomads are dancing.

You're aware of your weakness and your thirst. Your mouth is no longer dry, but you feel a thirst that stems from your core, out-ward through your limbs, and beyond them into the desert night, a thirst as broad as the Sahara itself. A veiled man now squats beside you and, his eyes glittering in the firelight over his veil, puts his hand behind your head to lift it slightly to a wooden bowl of water at your lips. You drink. It is the sweetest, coolest water you have ever tasted—artesian water that fell as rain far away on the Sahara in Paleolithic times and slowly migrated beneath the sands to seep from a break between the barren sandy hills and feed the palms of the oasis and their bunches of succulent, sun-sweetened dates. The water flowing between your lips seems a kind of miracle to you. This single bowl of water somehow survived amid the sucking dryness of the desert sands, the superheated, waterless atmosphere that hovers above it, and the desiccated vacuum of the heavens above that arches over both. All the gold and all the books in the world mean nothing be-side this one bowl of water. Without this single bowl of water, you are simply dust drifting through those heavens.

You drift off to sleep again to the sound of the drums and the

tambourines and the chanting, the veiled faces and glittering eyes
and swaying robes and bare breasts slowly wheeling in the firelight
against the now black sky. Are they doing all this chanting and moving on your behalf? You don't know. You don't really care at this
point. You lapse into dreams of valleys—dry, barren, sandy moonlit
valleys—one after the other that you are struggling across. Something good awaits you on the other side. You don't know why and
you don't know what, only that you must keep moving through the
valleys, down one sandy side, across the hard flat bottom, up the
other soft, bogging slope. You had in your hand a wooden bowl
that's extremely important, but somehow you've lost it. You miss it,
but you know you must leave it behind and keep going.

In your dream, you remember bits of the Rumi poem the Sufi
quoted for you:

> In whatever state you may be, keep on the search!
> Thou dry-lipped one, ever be on the search for water!
> Thou dry lip is a sure token
> That in the end it will find the source.
>
> This seeking is a blessed restlessness,
> It overcomes every obstacle and is the key to thy desires.
> Though thou have no vessel, fail not to seek . . .
> Sooner or later he who seeks become he who finds.

When you awaken again, the circle of faces has widened and is
moving faster, the drums and tambourines are louder, there is a
frenzy to the dance. A pair of dusty, bare, cracked, and callused feet
spin in the sand beside your head. You follow them upward to a
whirling patched robe, and then, illuminated by the flickering orange light, to a face that you recognize. It's the Sufi. He is spinning
rapidly in the center of the circle of nomads with his arms extended,
like a top. Now you understand what they are doing. He described
to you the Sufi practice of *sama*—a ritual of chanted poetry, much of

it love poetry, and music and dance that transports its participants into a trancelike state of ecstasy. In the center of the circle, the spinning Sufi represents the sun, and the dancers are the planets revolving around him.

But what is the Sufi doing here? He should be back at the last oasis, nearly two hundred miles across *reg* and *erg*.

One of the veiled nomad men from beyond the circle sees that you are awake again, and comes forward with a wooden bowl of water. He crouches beside you and once more puts it to your lips.

When you have drunk a few more sips, you try to address him through the drums and tambourines and chanting, using your few words of Tamachek.

"From where does this Sufi come?" you ask.

"From far away," says the nomad.

"But how did he get here?" you ask.

"Do you not know?" says the nomad. "It was he who brought you here. It was he who sent you out into the desert, and he who left his tracks for you to follow."

You will slowly come to understand other things, too, from your experience in the desert. The desert indeed is a place of nakedness that strips away the superfluous layers of the self as it peels away the fertile green layer of life that covers so much of the planet. Like the oxygenless summit of a Himalayan peak or the silent, motionless cold of a 40-below-zero Arctic night, the desert teaches you just how thin that layer of life is, and how fragile your own hold on it. Stepping beyond that fragile layer is no more difficult than shedding your clothes on a cold winter night or walking for a few hours without water in the hot sun. And when you finally do step beyond it, the ego, the vanity, the insignificance, and—often—the pettiness of so much of what passes for human endeavor and striving become abundantly clear. This is what the great religions and the shamans and the Sufi all are trying to tell you—to step beyond the self that blinds you, as the Sufi knows, having stripped away material wealth and worldly ambitions in his pursuit of union, spinning with ecstasy under the stars in the naked desert night.

This is why you climb mountains, and paddle whitewater rivers, and trek into the desert, and seek out remote places; to strip away the superfluous, to remove the protective boundaries between that thing you call a self and something larger. Your body still lies weakened and half shriveled, slowly taking on water to restore itself fully to life, but already as you lie in the sand with the flames of the fire jumping into the night, the desert tempts you back to it. The mountains tempt you to climb them. The rivers tempt you to run them. The remote places beckon.

It is not a question of whether you will go again.

The question for you is: where next?

And the other question is: how far?

SOURCES

Wilderness Medicine Instruction and Reference Books

Auerbach, Paul S., M.D., Howard J. Donner, M.D., and Eric A. Weiss, M.D. *Field Guide to Wilderness Medicine.* St. Louis: Mosby, 1999.

Auerbach, Paul S., M.D. (ed.). *Wilderness Medicine: Management of Wilderness and Environmental Emergencies.* St. Louis: Mosby, 1995.

Forgey, William, M.D. *Wilderness Medicine: Beyond First Aid.* Guilford, CT: Globe Pequot Press, 1999.

Wilkerson, James, M.D. *Medicine for Mountaineering and Other Wilderness Activities.* Seattle: The Mountaineers, 1992.

General

Bodanis, David. *The Body Book.* Boston: Little, Brown and Company, 1984.

Byock, Ira, M.D. *Dying Well: The Prospect for Growth at the End of Life.* New York: Riverhead Books, 1997.

Edholm, Dr. O. G., and A. L. Bacharach. *The Physiology of Human Survival.* London: Academic Press, 1975.

Editors of the National Geographic Society. *The Incredible Machine.* Washington, DC: National Geographic Society, 1986.

Hoffmann, Yoel (ed.). *Japanese Death Poems.* Rutland, VT: Charles E. Tuttle Company, 1986.

Isselbacher, Kurt J., M.D., et al. (eds.). *Harrison's Principles of Internal Medicine.* New York: McGraw-Hill, 1994.

Kübler-Ross, Elisabeth, M.D. *On Death and Dying.* New York: Simon and Schuster, 1969.

———. *Death: The Final Stage of Growth.* New York: Simon and Schuster, 1975.

Nuland, Sherwin B., M.D. *How We Die: Reflections on Life's Final Chapter.* New York: Alfred A. Knopf, 1994.

Pandolf, Kent B., Michael N. Sawka, and Richard R. Gonzalez (eds.). *Human Performance Physiology and Environmental Medicine at Terrestrial Extremes.* Carmel, IN: Cooper Publishing Group, n.d.

Schwartz, George R., M.D. *Principles and Practice of Emergency Medicine.* Philadelphia: Lippincott Williams and Wilkins, 1999.

Stein, Jay H., M.D. (ed.). *Internal Medicine.* St. Louis: Mosby, 1994.

Takahashi, Takeo. *Atlas of the Human Body.* New York: Harper-Collins, 1994.

Mountain Sickness

Blum, Arlene. *Annapurna: A Woman's Place.* San Francisco: Sierra Club Books, 1980.

Curran, Jim. *K2: Triumph and Tragedy.* Boston: Houghton Mifflin, 1987.

Davidson, Art. *Minus 148°: First Winter Ascent of Mt. McKinley.* Seattle: The Mountaineers, 1969.

Evans-Wentz, W. Y. *The Tibetan Book of the Dead.* New York: Oxford University Press, 1960.

Herzog, Maurice. *Annapurna.* New York: Lyons Press, 1997. (Originally published: New York: Dutton, 1952.)

Houston, Charles S., M.D. *Going High: The Story of Man and Altitude.* Burlington, VT: Charles S. Houston, M.D., and American Alpine Club, 1980.

Hultgren, Herbert N., M.D. *High Altitude Medicine.* Stanford, CA: Hultgren Publications, 1997.

Iserson, Kenneth V., M.D. *Death to Dust: What Happens to Dead Bodies?* Tucson, AZ: Galen Press, 1994.

Krakauer, Jon. *Into Thin Air.* New York: Villard, 1997.

Sogyal, Rinpoche. *The Tibetan Book of Living and Dying.* New York: HarperCollins, 1993.

Sutton, John R., Charles S. Houston, and Geoffrey Coates (eds.). *Hypoxia and Cold.* New York: Praeger Publishers, 1980.

Thurman, Robert A. F. (trans.). *The Tibetan Book of the Dead: Liberation Through Understanding in the Between.* New York: Bantam Books, 1994.

Tullis, Julie. *Clouds from Both Sides: An Autobiography.* San Francisco: Sierra Club Books, 1987.

Avalanche

Armstrong, Betsy, and Knox Williams. *The Avalanche Book,* Golden, CO: Fulcrum, 1986.

Carter, Rita. *Mapping the Mind.* Berkeley: University of California Press, 1998.

Kotulak, Ronald. *Inside the Brain.* Kansas City: Andrews and McMeel, 1996.

Ornstein, Robert, and Richard F. Thompson. *The Amazing Brain.* Boston: Houghton Mifflin, 1984.

Ramachandran, V. S., M.D., and Sandra Blakeslee. *Phantoms in the Brain: Probing the Mysteries of the Human Mind.* New York: William Morrow, 1998.

Robbins, Jim. *A Symphony in the Brain: The Evolution of the New Brain Wave Biofeedback.* New York: Atlantic Monthly Press, 2000.

Bends

Berger, Karen. *Scuba Diving.* New York: W. W. Norton, 2000.

Bove, Alfred A., M.D. (ed.). *Diving Medicine.* Philadelphia: Harcourt Brace, 1997.

De Latil, Pierre, and Jean Rivoire. *Man and the Underwater World.* New York: G. P. Putnam's Sons, 1956.

Lyon, Eugene. *The Search for the Atocha.* New York: Harper and Row, 1979.

Mathewson, R. Duncan III. *Treasure of the Atocha.* New York: E. P. Dutton, 1986.

Meyer, Franz O. *Diving and Snorkeling Guide to Belize.* Houston, TX: Gulf Publishing, 1990.

Middleton, Ned. *Diving Belize.* Locust Valley, NY: Aqua Quest Publications, 1994.

Dehydration

Adolph, E. F., and Associates. *Physiology of Man in the Desert.* New York: Interscience Publishers, 1947.

Bowles, Paul. *Their Heads Are Green and Their Hands Are Blue.* New York: Ecco Press, 1984.

Christopher, Robert, with Erik James Martin. *Ocean of Fire: From the Garden of Allah to Timbuktu.* New York: Rand McNally, 1956.

Englebert, Victor. *Wind, Sand and Silence: Travels with Africa's Last Nomads.* San Francisco: Chronicle Books, 1992.

Ernst, Carl W. *The Shambhala Guide to Sufism.* Boston: Shambhala Publications, 1997.

Gardi, René. *Sahara.* Bern, Switzerland: Kümmerly and Frey Geographical Publishers, 1970.

Gautier, E. F. *Sahara: The Great Desert.* New York: Columbia University Press, 1935.

Kruger, Christoph (ed.). *Sahara*. New York: G. P. Putnam's Sons, 1969.

Rice, Cyprian. *The Persian Sufis*. London: George Allen and Unwin, 1964.

Smith, Margaret. *Readings from the Mystics of Islam*. London: Luzac, 1972.

Von Dumreicher, André. *Trackers and Smugglers in the Deserts of Egypt*. New York: Dial Press, 1931.

Wolf, A. V. *Thirst: Physiology of the Urge to Drink and Problems of Water Lack*. Springfield, IL: Charles C. Thomas, 1958.

Drowning

Lao Tzu. *Tao Te Ching*. Trans. by Gia-Fu Feng and Jane English. New York: Random House, 1989.

Modell, Jerome H. *The Pathophysiology and Treatment of Drowning and Near-Drowning*. Springfield, IL: Charles C. Thomas, 1971.

Shepard, Roy J., M.D. *Exercise Physiology*. Toronto: B. C. Decker, 1987.

Strauss, Richard H., M.D. (ed.). *Sports Medicine and Physiology*. Philadelphia: W. B. Saunders, 1979.

Falling

Gardiner, Steve. *Why I Climb: Personal Insights of Top Climbers*. Harrisburg, PA: Stackpole Books, 1990.

Long, John. *How to Rock Climb*. Evergreen, CO: Chockstone Press, 1989.

Potterfield, Peter. *In the Zone: Epic Survival Stories from the Mountaineering World*. Seattle: The Mountaineers, 1996.

Poynter, Dan. *Parachuting: The Skydiver's Handbook*. Santa Barbara, CA: Para Publishing, 1992.

Roberts, David. *Moments of Doubt*. Seattle: The Mountaineers, 1986.

Simpson, Joe. *Touching the Void*. New York: Harper and Row, 1988.

Heatstroke

Edholm, Otto G. *Man—Hot and Cold*. London: Edward Arnold, 1978.

Ingram, D. L., and L. E. Mount. *Man and Animals in Hot Environments*. New York: Springer-Verlag, 1975.

Kavaler, Lucy. *A Matter of Degree: Heat, Life and Death*. New York: Harper and Row, 1981.

Hypothermia

Burton, Alan C., and Otto G. Edholm. *Man in a Cold Environment: Physiological and Pathological Effects of Exposure to Low Temperatures*. London: Arnold, 1955.

Collins, K. J. *Hypothermia: The Facts*. New York: Oxford University Press, 1983.

Forgey, William W., M.D. *Basic Essentials: Hypothermia*. Guilford, CT: Globe Pequot Press, 1999.

Kavaler, Lucy. *Freezing Point: Cold as a Matter of Life and Death*. John Day, 1970.

Jellyfish and Predators

Caras, Roger A. *Dangerous to Man*. Philadelphia: Chilton Books, 1964.

Halstead, Bruce W., M.D. *Poisonous and Venomous Marine Animals of the World*. Princeton, NJ: Darwin Press, 1978.

Ricciuti, Edward R. *Killer Animals*. New York: Walker, 1976.

Williamson, John A., et al. (eds.). *Venomous and Poisonous Marine Animals: A Medical and Biological Handbook*. Sydney: University of New South Wales Press, 1996.

Malaria

Knell, A. J. (ed.). *Malaria: Publication of the Tropical Programme of the Wellcome Trust*. Oxford: Oxford University Press, 1991.

Kreier, Julius P. (ed.). *Malaria*. 3 vols. New York: Academic Press, 1980.

Oakes, Stanley C. Jr. (ed.). *Malaria: Obstacles and Opportunities: A Report of the Committee for the Study of Malaria Prevention and Control*. Washington, DC: National Academy Press, 1991.

Scurvy

Brody, Tom. *Nutritional Biochemistry*. San Diego: Academic Press, 1999.

Carpenter, Kenneth J. *The History of Scurvy and Vitamin C*. Cambridge: Cambridge University Press, 1986.

Cuppage, Francis E. *James Cook and the Conquest of Scurvy*. Westport, CT: Greenwood Press, 1994.

Deutsch, Ronald M. *Realities of Nutrition*. Palo Alto, CA: Bull Publishing, 1976.

Kutsky, Roman J. *Handbook of Vitamins, Minerals and Hormones.* New York: Van Nostrand Reinhold, 1981.

Merck Service Bulletin. *Vitamin C.* Rahway, NJ: Merck, 1956.

Steller, Georg Wilhelm. *Journal of a Voyage with Bering, 1741–1742.* O. W. Frost (ed.) and trans. by Margritt A. Engel and O. W. Frost. Stanford: Stanford University Press, 1988.

Wentzler, Rich. *The Vitamin Book.* New York: St. Martin's Press, 1978.

ABOUT THE AUTHOR

PETER STARK is a longtime contributor to *Outside* whose work has also appeared in *Smithsonian, The New Yorker,* and many other publications. His article for *Outside,* "As Freezing Persons Recollect the Snow"—the inspiration for this book—was cited as a notable essay in *Best American Essays* (1997). He has been nominated for a National Magazine Award and has published a collection of essays, *Driving to Greenland.* He is also the editor of an anthology of writings about the Arctic, *Ring of Ice.* His assignments and travels have taken him to the Arctic, Tibet, Manchuria, West Africa, Irian Jaya, Iceland, and the Sahara Desert. He lives in Missoula, Montana, with his wife and two young children.